Science for Fun

Science for Fun

Dr.S.B.Asoka Dissanayake

Asokaplus

Asokaplus

Contents

6

Chapter 01

Prologue

I never had a good teacher in science.

All of them were mediocre.

Perhaps, there was no need for me to have a teacher.

Looking back, I never trusted my science teachers.

Come to think about it, I did not even trust my teachers in medicine.

Some of them were real bluffers.

They were failures in their own career of teaching of science.

Most of them were poor achievers, nothing to look upon and emulate.

They would have been bigger failures had they actually persevered with any scientific endeavor of their own.

Teaching was the lesser of an offense and we were the poor guinea pigs.

Then how come I become interested in science?

In actual fact, I never thought about it till I decided to write this book.

I may have had good traits.

1. Ability to Observe.

2. Formulate my own simple explanations to something complex.

3. Ability to revise my own explanations when new stream of thoughts enter my mind with newly observed facts of life.

The above three were enough to see whether the teacher was telling the truth or was in fact, bluffing.

I also had the canny habit of not letting them know what I was thinking inside with many ideas whirling inside my head.

I would please them with answers to suit their ignorance.

But out of sight and with my friends I would give my own appropriate explanations.

Unfortunately, I did not have similar talented kids in my class.

When I was fed up with a school, I would refuse to go to school and would ask my father to find a better one.

Unfortunately, he could not find a better school with better teachers.

Since, I did not have big ambitions, that did not matter to me at all.

I was reprimanded for being smart.

Finally, I decided to come out of the class and did it my own way.

Do it myself, the D.I.M, or the Dime Way.

I was further punished and I was not allowed to sit for medical but for biology, on the ground that I did not study enough organic chemistry.

In actual fact, we did not have a teacher to teach organic chemistry.

We had to do the organic chemistry practical in the university pre-entrance practical examination. So, I argued that I would do that after the ´A´ Level examination.

Then I was told the name has been already entered for biology and I need to sign in front of it.

I was forced to do that for a week and I flatly refused knowing very well, if I did not sign, the entire class will miss the examination. Then, finally the day before the closing day, I was forced again but I refused and went and reported the matter to the principal.

This was the principal who thrashed me when he was the vice principal with a stick till it broke into several pieces and I refused to budge and join the senior cadet team. Mind you, I was the captain of the junior cadet team in my previous school which came within the first 3 in all island competition at Diyathalawa.

He knew I was a tough nut and asked me would you be able to do that.

I said; a Big Yes Sir.

He called the teacher and asked him to enter my name for Medicine.

He abruptly said, there are no forms, I had to sign as was entered in the form.

Then the principal got frothing wild and told the teacher to send a telegram and get one before midday.

The principal took the telephone and made a personal request to the local Education Department, for an extra form.

I volunteered to walk to the Department and fetch it myself since I knew there will be only one name in that list.

I went there got a form and entered my name legibly signed it and checked it with the documentation officer to see whether it was alright and brought it back by 11 A.M. to the school.

By the time I reached the school the said class teacher had already sent the other application list and my one went by express post to the Department of Examination.

There was some thing pleasant, after all this.

On the examination day, I had one cubicle for myself and there were no other candidates. There had to be a separate examiner for me all throughout the examination and the news went around the examiners that this boy was a hard nut and never left the cubicle until I finished answering the paper.

All the papers came separately and the answer scripts were also packed separately.

During the examination, I told my friends, if I get through the examination, all these teachers except the principal have to leave the school by default or I will get them out by other means.

I had my own plan to get an Organic Chemistry Teacher and the Science School Inspector was more than helpful.

Sure enough, I was the only one successful in Medicine and I did very well in organic chemistry practical. One of specimens given to me did not react to any of the tests and I did not guess the compound and only reported as such non-reactive, most likely anthracene or similar compound.

I got the highest mark in that group, anyway.

That is the overt part of my genes.

But what were the hidden traits?

I can only guess.

It all dates back to preschool era.

I had seen the Sputnik in the sky.

I have heard about the Lyka, the bitch who was sent to the space and died in the process.

The nineteen fifties were full of science fiction.

Strangely enough, I only heard about Sir Arther C. Clark very late in my life and read only a few of his books but was more than interested in Scientific American, even in my advanced level stage and in the first few years of my career in the university.

Chapter 02

Scatologists

This is neither a singing scat (modern Sinhala variety) nor a literary exegesis. Nor any obscenity hidden or but a scientific exercise of forensic intent only. The courage, which I lacked in exposition of this exegesis, is rekindled after a casual glancing of a page in Scientific American magazine, which I had visited earlier but had failed to take notice of the particular page.

Scatology is an ancient art of our hunter gatherer revisited by the naturalists of the western hemisphere looking for vanishing species and their food fads. This particular lady scientist gather all the evidence of scat in the North American wasteland looking for the indigenous species and their habitat. She collects all the specimens with an enthusiasm, which our doctors lack when their patients come with an ailment of the type of "bada passa amma".

In other words gastro-colic symptoms of the human variety.

I was of the belief that I lost interest of this science of the human scat (looking for helminth eggs) in my internship. Looking back somewhat philosophically quite subconsciously I had been a scatologist all along, whether I was here in this country or abroad.

In England of course they should have a degree course in scatology with forensic expertise to see who is dirtying the footpath to one's home. It is a perennial unsolved problem in England for that matter Australia and New Zealand too.

I have seen a duel between a doctor who remains in this country and a doctor who had gone abroad and having come on a holiday advocating killing all Kandy stray dogs.

I prescribe the study of scatology for the doctor who had come from abroad as my simple medical advice (finding a way of cleaning London's residential streets).

Please leave our innocent stray dogs as they are since they are the watchdogs of our nature and the habitat.

This Sri-Lankan should be mindful to the fact that the dogs were sacrificed during the Perehara season and quite to the liking of the Koreans.

They were served as a delicacy.

Coming back to my subconscious interest in the scatology, I, in fact named our private footpath as BKL for short, "The Balu Kakka Lane" when I returned from abroad to the delight of my son.

He was a scatologist of a different kind.

He was of the dinosaur's variety and "Dino-Dropping" was his childhood interest. Walking down our lane one has to be almost in meditative mode with an attention to every footstep down the lane not the mind.

If not, one's foot is dirtied and with it the meditative mode of mind!

I follow the simple Dr. Adhicaram's advice on issues like that.

Avoid the danger diligently and get on with my mind's attention.

My interest in scatology was aroused recently by the loss of our cat during Perahara episode with some forensic interest.

For nearly two weeks I studied all the scats of the foreign variety.

In fact, the natural variety of Indian civet (Kalawaddha) and the Ceylon civet (Uruleva) who visit our neighbourhood, almost on daily basis. I never found any evidence of meat or undigested skin of our cat in the droppings. I always found only fruit variety of scat and never the meat variety even though I believed they were coming for the domestic rats.

The diversion of my attention to the Kandy Municipality of course came after my initial investigations. They were killing the stray dogs (poisoning) and cats and our cat of course, one of the victims.

My form conviction was some of them ended up as substitute for chicken in some hotels during Perehara festival.

It did not bother me in anyway since, I had given up eating any meat long before.

Alas, I did not have the luxury of examining the human scat on this occasion. I think one should revisit the ancient habit of going after the dropping for a hunt.

The amount of commercial and political Aggalas we had during the recent election it is time for taking stocks of the scats.

A degree course is warranted for the politicians and not the naturalists.

Chapter 03

Back to Basics

Everybody in the family out of the country on a short holiday, I was getting ready for a quiet time at home with cricket as a top priority for entertainment and bit of writing in between, if I felt like or hit by a brain storm.

All my preparation went awry.

Who is to be blamed?

I had strict instruction from my daughter that I should look after the dog during their absence and nothing should go wrong.

That is a piece of cake, I thought to myself.

I was caught wrong footed by our pet dog well and truly. He was all right on the first night. Only spilled the milk, as a form of protest but that was only the beginning of series of teenage temper tantrums. For a week I was trying to get him to drink the milk but failed. Almost half a packet of powdered milk wasted but I was about to give up.

Now there is no cat at home, I could not offer what was leftover, or the milk to the cat. The milk spilled all over the house by our teenage (3 years exactly) rascal made me to revise my strategy.

By Sunday I found a way out but I would come to that point later.

First of all I should canonize all the mistakes I did.

Like a typical Sri-Lankan, I took the easy way out rather the lazy way out on the first evening. On my way home I went to one of the Chinese restaurants and bought a fish grill with fried rice and returned home.

The fish portion was fairly large and half of it was put in the fridge (I did not eat any fish thinking that it would save another journey to the restaurant) and I had rice with leftover curries.

The lunch was OK.

Fish fingers and chips were our dinner.

The dog was happy but problem was with the milk, which was spilled with his snout.

Next day, breakfast fasted, lunch refused and for dinner he ate only the pieces of fish.

The dog was on a hunger strike by now knowing that the entire family except me is out.

According to Kevin's expression I made my entire family disappear (classic film of Home Alone) and I could not tell the dog how I did that.

He was watching every movement of me and wanted an answer.

The next day he vented his anger on the biggest of the three Japanese Spitzs next door. He was chained with chain in my hand but I could not restrain. Quite out of the ordinary, luckily I had the chain in my hand, I could separate him, after a minute of struggling with some effort.

Next he took on the cat who had come to eat the food (refused food) that I had left outside.

I wondered where he gets his energy without even eating a mouthful.

The wild dogs perhaps can go even for a week without food but domestic ones perhaps not, I thought.

Looking at stray dogs on rubbish dumps that is the only logical impression I have. Hunger strike I could take but I was not ready for the inbuilt aggression.

Mistakes

1. I took the easy option of going for junk food.

2. I changed his routine without any planning.

3. I pressed the panic button.

4. Tried to overcompensate i.e. did not allow the dog to come to term with the sudden change in home anatomy.

5. Left the dog alone in desperation;

i.e. I did have to go to work and return home late.

Back to basics

Generally speaking the home food had a bias on vegetarian base. Within a week with some careful planning I went from fish base to vegetarian base.

By the next Sunday we were back to base.

Entire meals for Sunday were vegetarian and we did not have a problem.

It was with some relief now that I pen down this.

No bread (I hate bread) and eggs for a week.

The diet varied from chickpeas to mung beans to manioc to soybeans with gradual tailing off the fish base for the dog.

Beans left for sprouting, grated coconut (desiccated coconut, I bought was not used) cheese, yogurt and home made pizza did the trick.

Soya with Ambul thial did the final change over.

We were back to base and when the family returns, I can take a back seat in the kitchen environment.

I did not have to open the Samaposha for two weeks and that was also a record.

Only two cans of fish for the entire period, no noodles and eggs at all. Only supplement I used was a bit of Marmite paste, which my dog loved.

The biggest problem was milk (powdered).

He refused it for a week.

But I found a solution for that too.

The fresh cow's milk substitution solved the problem.

Even the dogs do not like artificial food whether it is milk powder or desiccated coconut and the lessons of the week was plenty of natural food and lot of time and common sense.

What was remarkable was the amount of rubbish was negligible, the tea leaves and the ash of the mosquito coils were saved for the indoor pots. (I was doing a private study of the effect of mosquito coils on indoor plants.)

The sodium channel blocker, I believe may have some effect on the plant cells. I have nothing concrete yet to record probably because the tea leaves offsetting the deleterious effects. I can recommend tea leaves for the indoor plants but one has to manage the stain it leaves on the floor.

Chapter 04

Butterfly Effect

I am one who vents

An air of proclamation, that

When a few butterflies

Flutter their wings

On my side of the world

Would ignite a chain reaction

On the other side of the world

In the far corners of the globe

In the Arctic or the Antarctic

Come it may be

In philosophical sense

With a scientific slant

And in an investigative mind

But not to stir up

The potential political undertones

But when I ask myself the question

How clean the clean air is?

And when I get the answer

That there are 1000 minute particles

Of dust in every cubic meter density

On this planet
But not in the clear looking Iceland
But the Antarctic
Gets the credit for
Such an accolade
In this contemporary world

When somebody ask me
Are you asking the right question?
Lest you get the wrong answer
Of some sort
By the Coal Power Merchants
Of the modern world
Who continue to discover
The latest technology

How clean my air and breath
Or how to dispose
The vast amount of ashes and sediment
That get deposited
On this tiny Island Paradise
With the power plant in full swing?

One who was brought up
With the tradition of zero tolerance
In life and matters of discordance
And purity par excellence
Having forced to go
From macro to micro to nano-tolerance
Of the 21st century macroeconomics
Driving me mad
Thinking about
 Going to the outer space,
For the safety limits of tolerance
For the Dead Space of my lung
But not the Vital Capacity
Which is fast receding
With time and age

But when I think
Of the virulent fungi
That grow endlessly
In the closed tubes
Of the artificial respiratory tracts
Of the Space Station
I wonder

I could be born again
In another Planet
Outside this Solar System
As an alien being of some sort
Bringing the alien wisdom
To this Island Paradise

Is it a wishful thinking of mine?

Chapter 05

Dearth of a Philosophy

Science with philosophy

And philosophy with science

Are both interdependent

On each other

Without which

Like a fish out of water

Kills the other

In its inception

Rather

The conception of some kind

By the very sentiment

Of dominance, deviance and the separation

Of existence in the

Modern

Way of life

Science and philosophy

Are two sides of the same coin

Quite inseparable

For their sustenance and subsistence

But

As a toss of a coin can land

On one side at a time

Masquerading each other

Not by intent

But by the very origin of

Its statistical design

As of husbands and wives

They have contrasting styles

In expressions and manifestations

From which originate

The essence and principles

Of good governance

But

Like the birth of a being

Into existence

The mother take precedence

Over the father

Just for the survival of the being

Concepts and ideas

In Natural Science

In general
The killing of the hen
Who laid the golden eggs
The purity of philosophy
Is diluted
By plethora of Opinions
Is evidently manifesting
And is truly unwelcome

Just like the toss of a coin
And its landing one one side
Is one to one or 50%
By chance
Variance of occurrences

But quite discriminatingly
And obtrusively
Science lands
Nine out of ten times
Or 90% of the time
Disturbingly high logic
Of the modern day and age
Of natural justice,

Fair play

And law of evidence

I think the coin should land on its

Thinner side

And should roll on regardless

With freedom

Without friction

Like the muscle that power

The bicycle

Move faster

Than the running soul

The invention of the wheel

Lead to the speed

And progress with efficiency

One need not carry

The wheel on one's head

Or on shoulders

But at ground level

Touching the essence

Of philosophy

Firmly grounded

And closely to the heart

For the proper survival
For another generation

The decadence of the 21st Century
Set in motion now
Is the result of
The imbalance
Of guiding principles
From which originate
The moral and ethical
Fabric and forbearance
Of modern civilization

One oblique reference
Is timely and relevant
To the origin of religions
Where the ethical principles and philosophy
Laid the foundation and structure

How come
All the modern religions
Appoint
Their dignitaries without exceptions

With abundance of gray hair
On the top?

At the present time
With Buddhist monks
Wearing
A hair style of reference
Not excluded

They have little gray matter
Remaining in the brain due
To age old atrophy

"Wearing a thinking cap"
Literally and metaphorically
Is grossly inadequate
Known by the
Utterances of pontifications
Disregarding wiser council
Of wisdom
And the very existence of
The Mankind threatened
By default or by design

With ulterior motives
A catalyst for perpetuating
Disdain, discordance
And violence of all kinds
Including verbal
Intellectual and physical
Be that it may be Jihads, Christianity
Or Atheism

How can a holy war be prescribed
For lack of prudence and forethought
Of this aging generations
For the ethical vacuum
Created by Science
Void of Religion
Divorced from
Philosophy of Mind?

That is the million dollar question
Of mine
To the vested interests of
The East and West
Who procrastinate

Didactic dogmas

Sans

Sense and Sensibility

No offense intended

Philosophically to any soul

On this mother Earth.

Chapter 06

Science

Just as there are
Many notions
To the origin of the Universe
There are many misconceptions
To the meaning of science
Which literally means
Acquisition of knowledge

The first of the many misconceptions
And the foremost of it is
That science is infallible
And the second that originates from the first
Is that science is the only explanation
From that derive the notion
That all interpretations
Are logical, complete and final

None of the above statements are true
In scientific sense
And that is the beauty of the science

Which is expanding

Challenging

And changing

By the day

Chapter 07

Science In Action

Scientific Jargon

There are a lot of jargon words to describe a scientific phenomenon. That is if one thinks of the phenomenon, only in pure scientific terms.

But one does not need all these scientific words if the thing that you observe is a living thing as a whole.

If it is a flower we give a name to it.

If it is an animal we give an appropriate name.

The one who is not inclined in scientific thinking that is all that is needed. He uses his natural language to describe what he sees and is happy about it.

He does not bother to describe how many petals there are and how they are arranged in a whorl?

But if one wants to describe the flower its anatomy and its function, for the convenience of the reader certain standard words have to be coined.

That is all what happens in scientific notation.

Nothing more and nothing less.

So, the number of jargon words increases with the diversity of living things around us and depending on the complexity of the structure that has to be described.

So one should not be frightened by the new word or words.

It is more fun than learning a second language.

But the difficulty is when one studies in a particular language, especially in a Eastern language where the scientific terminology was slow to integrate within its vocabulary, one has to coin a new word, not by design but by default since there is already word in English, French or German.

This new word or the translation sometimes become a handicap in learning.

That is why I encourage the use of English, since it is not a difficult language to learn.

But my interest in science, especially in biology developed naturally.

It was all because of a little beautiful fish called guppy fish.

The name the guppy comes from Dr. Robert Guppy (see below) who sent a specimen to London from Trinidad.

The female fish is unremarkable, gray in colour but the male has a beautiful and colourful tail. There is also blond variety due to mutation.

In fact, no two male guppy fishes, look the same. That was what made me interested and in later years in genetics and pigmentation in animal kingdom.

The flowers did not interest me a lot but I had to learn them as part of biology.

Now my main interest is water plants, and their flowers and lately water lilies.

So knowing a bit of science is important , if you want to become a gardener.

Below I have given few jargon words related to a flower and the guppy fish.

The description and naming in science is called taxonomy.

Flower

The flower is the reproductive unit of some plants (angiosperms). Parts of the flower include petals, sepals, one or more carpels (the female reproductive organs), and stamens (the male reproductive organs).

The Female Reproductive Organs

The pistil is the collective term for the carpel(s). Each carpel includes an ovary (where the ovules are produced; ovules are the female reproductive cells, the eggs), a style (a tube on top of the ovary), and a stigma (which receives the pollen during fertilization).

The Male Reproductive Organs

Stamens are the male reproductive parts of flowers. A stamen consists of an anther (which produces pollen) and a filament. The pollen consists of the male reproductive cells; they fertilize ovules.

Fertilization

Pollen must fertilize an ovule to produce a viable seed. This process is called pollination, and is often aided by insects, like bees, which fly from flower to flower collecting sweet nectar. As they visit flowers, they spread pollen around, depositing on some stigmas.

After a male pollen grain has landed on the stigma during fertilization, pollen tubes develop within the style, burrowing down to the ovary, where the sperm fertilizes an ovum (an egg cell), in the ovule.

After fertilization, the ovule develops into a seed in the ovary.

Types of Flowers

Some flowers (called perfect flowers) have both male and female reproductive organs; some flowers (called imperfect flowers) have only male reproductive organs or only female reproductive organs. Some plants have both male and female flowers, while other have males on one plant and females on another.

Complete flowers have stamens, a pistil, petals, and sepals.

Incomplete flowers lack one of these parts.

Male Parts
Stamen

This is the male part of the flower. It is made up of the filament and anther, it is the pollen producing part of the plant. The number of stamen is usually the same as the number of petals.

Anther

This is the part of the stamen that produces and contains pollen. It is usually on top of a long stalk that looks like a fine hair.

Filament

This is the fine hair-like stalk that the anther sits on top of.

Female Parts

Pistil

This is the female part of the flower. It is made up of the stigma, style, and ovary. Each pistil is constructed of one to many rolled leaf like structures.

Stigma

One of the female parts of the flower. It is the sticky bulb that you see in the center of the flowers, it is the part of the pistil of a flower which receives the pollen grains and on which they germinate.

Style

Another female part of the flower. This is the long stalk that the stigma sits on top of.

Ovary

The part of the plant, usually at the bottom of the flower, that has the seeds inside and turns into the fruit that we eat. The ovary contains ovules.

Ovule

The part of the ovary that becomes the seeds.

Other Important Parts of a Flower

Petal

The colorful, often bright part of the flower. They attract pollinators and are usually the reason why we buy and enjoy flowers.

Sepal

The parts that look like little green leaves that cover the outside of a flower bud to protect the flower before it opens.

Flower Types

Imperfect Flower

A flower that has either all male parts or all female parts, but not both in the same flower. Examples: cucumbers, pumpkin, and melons.

Perfect Flower

A flower that has both the male parts and female parts in the same flower.

Examples: roses, lilies, and dandelion.

Guppy Fish

The guppy (*Poecilia reticulata*), also known as million fish and rainbow fish, is one of the world's most widely distributed tropical fish, and one of the most popular freshwater aquarium fish species.

It is a member of the Poeciliidae family and, like all other members of the family, is live bearing.

Guppies, whose natural range is in northeast of South America, were introduced to many habitats and are now found all over the world. They are highly adaptable and thrive in many different environmental and ecological conditions.

Male guppies, which are smaller than females, have ornamental caudal and dorsal fins, while females are duller in colour.

Wild guppies generally feed on a variety of food sources, including algae and aquatic insect larvae, especially mosquito larva.

Guppies are used as a model organism in the field of ecology, evolution, and behavioural studies.

Taxonomy

Guppies were first described in Venezuela as *Poecilia reticulata* by Wilhelm Peters in 1859 and as *Lebistes poecilioides* in Barbados by De Filippi in 1861. It was named *Girardinus guppii* by Albert Günther in honor of Robert John Lechmere Guppy, who sent specimens of the species from Trinidad to the Natural History Museum in London.

It was reclassified as *Lebistes reticulatus* by Regan in 1913. Then in 1963, Rosen and Bailey brought it back to its original name, *Poecilia reticulata.*

While the taxonomy of the species was frequently changed and resulted in many synonyms, "guppy" remains the common name.

Distribution and habitat

Guppies are native to Antigua and Barbuda, Barbados, Brazil, Guyana, Jamaica,

Trinidad and Tobago, the U.S. Virgin Islands, and Venezuela.

However, guppies have been introduced to many different countries on every continent except Antarctica.

Sometimes this has occurred accidentally, but most often as a means of mosquito control. The guppies were expected to eat the mosquito larvae and help slow the spread of malaria.

Field studies reveal that guppies have colonized almost every freshwater body accessible to them in their natural ranges, especially in the streams located near the coastal fringes of mainland South America.

They tend to be more abundant in smaller streams and pools than in large, deep, or fast flowing rivers.

Guppy breeds

Guppies exhibit sexual dimorphism. While wild type females are gray in body color, males have splashes, spots, or stripes that can be any of a wide variety of colors.

The size of guppies vary, but males are typically 1.5–3.5 cm (0.6–1.4 in) long, while females are 3–6 cm (1.2–2.4 in) long.

A variety of guppy strains are produced by breeders through selective breeding, characterized by different colours, patterns, shapes, and sizes of fins, such as snake skin.

Many domestic strains have morphological traits that are very distinct from the wild type antecedents.

Males and females of many domestic strains usually have larger body size and are much more lavishly ornamented than their wild type antecedents.

Guppies have 23 pairs of chromosomes, including one pair of sex chromosomes, the same number as humans. The genes responsible for male guppies' ornamentations are male chromosome Y linked and are heritable.

Chapter 08

Teaching Science-01

Approach to teaching science is changing in the West. It is realized that in spite of the rapid advances in science (in the past century) the benefit of science has not filtered down to the masses, specially to the young (in and out of school).

When out of school the young adult has no way of furthering his or her basic learning in science for one's own benefit functionally and socially. The school should be a place to make young people learn in a constructive way (not irrelevant factual knowledge) and apply that knowledge with common sense and with a scientific approach.

There are inherent religious, ethnic and educational prejudices that one has to circumvent to achieve these goals without upsetting the social sensitivities.

Young children are brought up in various ways according to the way of life they are born to.

Some are enriched and others are deprived of valuable opportunity due to many factors beyond the control of the child. The school should be the ideal place that can change this social disadvantage and teaching science with a new and scientific way is desirable and is found to be wanting.

The art of teaching science has to change form what it is now in our schools.

The top down approach and the assumption that the teacher as a sage, knows all what the students should know has failed in this country and elsewhere whether it is the East or the West. In the West however, there is a changing attitude to teaching science (in a more pragmatic way).

My attempt here is to look at this problem in a child's perspective rather than the teachers perspective. It is not possible for a teacher to know (for that matter even a doctor to know all the aspects of medicine) all the aspects of science.

A mixture of integrative approach and pooling of resources rather than training pure science teachers is what is needed.

Instead of a purely scientific and academic exercise it should be broad based taking the child's development stage and the social background.

The medium of instruction and the command in whatever the language that is utilized to teach is also important. Bombarding factual knowledge to young is not what is required.

What should be tested or trained is not the memory and factual contents but the way of thinking based on scientific reasoning.

It is the thinking capacity that should be developed not the retention capacity.

Even though, I attempt this at the university level the outcome is not as it is expected by me in general.

What is required is to make students ready for acquiring concepts appropriate for their psychological development. Children vary in their stage of development not only in the capacity but also at what level they achieve certain milestones of conceptual development.

Brief description of the psychological development is necessary before discussing the Outcome Based Education (OBE).

One should not expect the children to grow developmentally in a rigid or programmed pattern. Even though, they follow a general pattern some are slow and some are fast in acquiring conceptual skills. It is a normal behaviour of conceptual development and it is not that a particular child is stupid or very smart.

It is the way they are born with and the teachers task is for the whole class to achieve a certain level of competence (achievement) in thinking science.

To cover a particular syllabus forced on them by the hierarchy in the education department who does not have a good understanding (but just follow what was prescribed for the past 40 years without any revision) of the current trends and thinking in education.

Changing requirements of the higher education should be balanced by the new way of thinking and approach in teaching science.

From concrete operational (direct experience) thinking to formal (acceptable to science) operational thinking is a very big step in psychological development. This change occur around 5 to 7 years of age and go on until 16 years of age. What is amazing is that children (and many adults too) do not shed their direct experience (and their thinking with it) when the conceptual development is progressing in their development (which has inherent variability).

This is a handicap (in a way blessing in disguise) the children from the age of 8 to 16 years undergo and overcome. Bombarding with factual knowledge at this stage of development is a serious problem from the point of view of the child. It is believed that conceptual development occurs at around the age of 16 and many do not develop this even up to 30 years of age and even then the conceptual development function at a very rudimentary level.

There is another barrier that is associated with psychological development. It is the language barrier itself. The language and grammatical structure did not evolve in parallel with science and its development. Language development and acquisition pre-dated the development of science by more than 2000 years.

It is only in the last hundred years that the Language of Science started integrating with the standard languages of the West. Scientific terms that emerged in the West were not kind to Eastern languages and if not for the Pali Language (specially Abhidhamma terminology) teaching science in Sinhala would have been much more difficult. Unfortunately the people who started translating scientific terminology went in tangents to the tenet of Pali which was meant for a different purpose.

By doing this they not only destroyed Sinhala but also the teaching of science by distorted interpretation of an ancient language.

The end result is that we are producing poor quality scientific thinkers and educators. I am one who believe that science should be taught in English very early in children's education. Not only this will help improve the English that the children would learn to use and would be an advantage in grasping difficult concepts in later life (especially when they proceed to higher studies).

Having seen students struggling to understand simple concepts in their first year at the university and the introduction English language learning before commencing their academic career has not improved or remedied the problem as it is, I am more than convinced now to say, English is the way forward for learning science.

My current impressions are somewhat biased and the reader should excuse me for that.

Having said that English is not a difficult language to learn and English is a very good supplement even if one wants to learn purely in one's own mother tongue.

Conceptual Development

Modeling is an acceptable way of expressing situations which we cannot observe directly. Models help us to understand what we observe and predict what will happen in situations that we cannot observe. Modeling a scientific concept is an acceptable scientific tool in advancing a concept or an idea. Eventually we may become so convinced of the reliability of the model or the security of the model image. Then we tend to accept it as total reality without any objection or question. The distinction between the model and the reality disappears in our teaching of science and this in turn inhibits the acceptance of new ideas.

The atomic model is a good example and everybody accepts it as a good model and almost certain reality but few thinkers deviated from this mode of thinking and went into talk about quarks and antiparticles and the like.

Our educationists in science have gone into hibernation after the atomic model and canceling the practicals at the university entrance for the advance level examination (due to inherent inability to hold examinations full proof so that some elements can doctor the results) compounded the already deteriorating approach to teaching science in schools.

This is one example of, the skewed (squint) eye view of teaching science and education in general.

We have let this phenomenon (science teaching without experiment and investigation at least at a rudimentary level) go on, for nearly 40 years without a review.

It is high time at least from "middle school age" the children are brought up looking for both knowledge and wisdom with open mind rather than "programed minds" of the education department. Restricting them to free books published by the education department which are probably 30 years outdated is not the panacea for a long term problem of "learning on the go" especially in science.

Australian teachers researching in science teaching are introducing many models of art of teaching science and I would like to jot down some thoughts about one of them in some detail with my own adaptation (not adoption) of that model.

5E Model

It is called the 5E Model (I am quite at ease with the Japanese 5S model of Quality Training) and worth some elaboration.

Students acquire a certain level of conceptual development on a scientific phenomenon that is discussed based on 5E Model.

The focus is on a observable topic or a phenomenon.

Engage

Students express their views and grasp of a particular phenomenon (or a topic) and study any connection with what they already seem to know at a particular age of their development. The teacher generate the new idea or the concept without any factual information.

Explore

Students explore the phenomenon and use their own language to express and discuss the phenomenon in their own understanding and experience without any hindrance from the teacher who act as a facilitator rather than a disseminator (discriminator) of knowledge. Students generate new ideas based on hands on activities.

Explain

Then, explain the phenomenon in scientific terms and teacher helps to develop new ideas, concepts and terminology.

Elaborate

The teacher introduces new challenging situation and students apply what they have learnt to a new situation and modify their concepts accordingly. This is the application of the newly acquired knowledge to a given problem. Problem oriented but evidenced based learning is gradually introduced and facilitated.

Evaluation

The students reflect on the preconceived views (ideas) to the post conceptual development of the new scientific theme and teacher allows ongoing development of the theme as and when the experience of the class improves with new situations and problems.

Building on a sound base of conceptualization children learn and understand more complex phenomena.

A layer of conceptual (idea) development is formed on top of the already existing concrete operational thinking which each student inherit differently without upsetting their own previous preconceived ideas and thinking capacities. There is no label attached to the previous conceived idea as either right or wrong but a modified version of the previous concept is laid on top the previous experience which each student inherited with their own experience or lack of it.

The thinking capacity is developed instead of retaining capacity.

Unfortunately there is tremendous variation of memory capacity of children.

The top or the surface (outermost) layer is kept open ended for further ongoing improvement and development of new ideas and concepts.

The approach is not rigid as it is in the book oriented learning expectation of the department of education.

In effect there are there layers of conceptual development.

Layer 1

Students already perceived thinking and ideas.

Layer 2

The new concepts developed with exploration and explanation.

Layer 3

Open end for further facilitation with new experience and understanding.

The third layer is kept open in true scientific spirits.

What is understood is kept for further development and refinement.

With this level of understanding and maturity students could enter what ever the type of further education they wish in their life. It should not be a static and should not end up abruptly at advanced level examination. The concept of continuous learning experience should be ingrained in the minds of school leavers who are encouraged to learn in their own pace and time rather than regimented time scale as prescribed by the education department.

Before elaborating with some examples that can be used in the class room I should introduce few of my own 5Es to make the grand total ten.

The are;

1) Evidence based,

2) Expansion of the,

3) Experience of the individual to,

4) Encourage true

5) Emancipation of knowledge,

Leading to right thinking.

Whatever the field of study one is free to apply this framework of exploration.

My intentions are to promote open discussion.

Chapter 09

Teaching Science-02

Weather Model
For Scientific Investigation

As an example the weather can be investigated in a scientific way. This examination is not complete or comprehensive but taken as an instruction model for discussion and refinement. The limitations inherent in any model in illustrating a scientific fact (as a form of analogy) is also discussed briefly.

Whether one studies mathematics, chemistry or physics, science teachers are very comfortable in using equations to explain scientific themes. These equations have well balanced structures and are ideal for explanation (not investigation) of scientific notions but may fall short of the reality.

That is something I would like to delve into.

My intention is to highlight the fallacy of using equations and equilibriums to state scientific facts as absolute truths (in a philosophical sense – the meaning of the word meaning?).

This approach of equations and equilibrium fails miserably when discussing the pattern of weather and its behaviour.

This was what I had encountered as a child and still do and the tsunami was an eye opener to rekindle my latent interests in science of weather reporting.

Even though, this is not and attempt or in depth analysis of weather or its reporting, an oblique reference is made to weather as a focal tenet of disagreement with the way the science is taught in our class rooms.

I have started addressing this in the Part 1 of the writing and this should be read in continuity with that and the Part 3 of my own observations that follow.

In weather an equilibrium state is never achieved in a scientific sense.

It illustrates the uncertainty principle in general.

Instead of an equilibrium state, cyclic phenomena are evident in weather patterns.

Weather is discussed in some detail below. The order and change (chaos) can be grasped without any difficulty unlike other scientific principles. The cyclical nature is apparent when one talks about monsoons and inter-monsoonal rain. What is evident is constant change but the order only becomes apparent because of repeated sequence of events (change and cyclic change adequately fits into this notion).

The cyclic nature of the phenomena is studied when forecasting of weather and it behaviour.

The model is the cyclic pattern of the water cycle.

Discussion is based on a scientific description of clouds and their behaviour which I copied from a web site with modification to suit the current discussion.

To begin with there were over a hundred new scientific and technical words in a short space of few passages. That is what I consider as the biggest handicap to a student or an average inquirer with open mind.

I would attempt to put that in perspective in simple terms as is possible but there is no guarantee.

It is a bold attempt since the few passages that I cover involve the entire package of scientific domains form physics to chemistry to mathematics to dynamics and logic and logistics.

Building Clouds

Although the formation of clouds can be quite complex in full detail, it can be simplified for a wider nonscientific audience.

There are two basic ingredients to satisfy formation of clouds;

Water and Dust.

On earth naturally occurring clouds are composed of either water in its liquid or solid state.

On other planets, where the surface atmosphere is different from that of the planet earth clouds may form from other compounds and that is not under discussion, here.

Thus, the primary recipe in forming clouds is water.

Collection of a sufficient quantity of water in a given space in its vapour state at a given time when the essential prerequisites (the temperature, the altitude, the pressure and the movement of molecules) are met the water vapour is transformed into clouds in either liquid or solid state.

The water vapour content of the atmosphere varies from near zero to about 100 percent, depending on the moisture on the surface beneath and the air temperature and condensation.

The water vapour content at a certain point of time and space needs to be ideally saturated to form clouds.

Next, recipe one needs is some dust.

Without "dirty air" there would be no clouds at all or only at high altitude consisting of ice (crystals) clouds. Earth atmosphere is never clean as one would expect it to be for healthy living (man's perspective).

Even the "cleanest" air found to contain about 1000 dust particles per cubic meter of air.

Neither a large amount nor the size of the particles nor all dusts would satisfy the primary needs of forming clouds.

Dust is needed as condensation (nidus or the nucleus) sites on which water vapour may condense or deposit as a water droplets (liquid) or ice crystals (solid). Certain types and shapes of dust and salt particles, such as sea salts and clay, make the best condensation nuclei. With proper quantities of water vapour and dust in an air parcel, the next step that has to be satisfied is the cooling of that air mass (i.e; cooling of the air parcel having a dust content of particular size and shape) to a particular temperature conducive for the formation of cloud droplets or ice crystals (suspended in air in as a massive aggregation).

Viola, the clouds are formed.

Just as there are many ways to prepare a recipe, there are many different ways to form clouds. The recipe can be expanded with new ingredients for the precipitation to occur.

Professor John Day, the Cloud Man, has taken the simple cloud recipe, added a few more details and continued it until it makes precipitation (rain).

He calls this 'The Precipitation Ladder.'

As with a simple recipe, he begins the process with the basic ingredients of dirty air and water vapour. As with cooking it is regulated (in real sense there is no regulators but changing states) to achieve the desired effect.

It is actually the opposite of cooking, the cooling (effect) of water vapour that initiate the cloud formation.

He takes the ingredients through the rungs of the ladder in several stages (several processes) to form a cloud.

Ascent and Expansion are two of the main processes that result in the cooling of an air parcel in which clouds will form. We mostly think of moving air as wind flowing horizontally across the surface (The movement of air is almost chaotic in a scientific sense but it is not completely at random).

Air moving vertically is extremely important in weather processes, particularly with respect to clouds and precipitation. Ascending air currents takes the process up the Precipitation Ladder.

The processes are assumed to be reversible.

With descending air currents the process comes down the ladder reversing the effects until finally water vapour and dust are left in the air stream (mass of atmosphere) of movement.

There are four main processes occurring at or near the earth's surface which give rise to convergence, convection, frontal lifting and physical lifting of the ascending air.

Convergence occurs when several surface air currents in the horizontal flow move toward each other to meet in a common front. When they converge, there is only one way to go and it is upwards only.

An area of low pressure (cell) build up, on the surface of the earth is an example of where the converging air currents result in rising of air at the center of the converging currents. The air at the center rises to accommodate the redistribution of various air pressures (wind) that build up due to variable degree of cooling and warming of the atmospheric air creating low and high pressure points in the atmosphere.

Convection occurs when air is heated by contact with a warmer land surface until it becomes less dense than the air above it. The heated parcel of air will rise until it has again cooled to the temperature of the surrounding air.

Frontal lifting occurs when a warmer air mass meets a colder one. Since warm air is less dense than cold, a warm air mass approaching a cold one will ascend over the cold air.

This forms a warm front.

When a cold air mass approaches a warm one, it wedges under the warmer air, lifting it above the ground.

This forms a cold front.

In either case, there is ascending air at the frontal boundary.

Physical lifting, also known as orographic lifting, occurs when horizontal winds are forced to rise in order to cross topographical barriers such as hills and mountains. Whatever the process causing an air parcel (volume or quantity) to ascend, the result is that the rising air parcel must change its pressure to be in equilibrium with the surrounding air. Since atmospheric pressure decreases with altitude, so too must the pressure of the ascending air parcel.

As air ascends, it expands.

And as it expands, it cools.

And the higher the parcel rises, the cooler it becomes.

Now that the cooling has begun the air parcel is almost ready to form a cloud.

The air parcel cools until condensation point is reached.

The next several rungs of the Precipitation Ladder describe the processes through to the condensation of liquid water.

As the air cools, its relative humidity increases, a process Prof. Day terms humidification.

Although nothing has yet happened to change the water vapour content of the air, the saturation threshold of the air parcel decreases as the air becomes cooler. With decreasing saturation threshold the relative humidity increases proportionately.

Cooling is the most important method for increasing the relative humidity but it is not the only one.

Another is to receive more water vapour through evaporation or mixing with humid air that come in contact (cloud that has already formed) with the result of moving air currents containing more (various degrees) water vapour.

To form a cloud, humidification may eventually bring the air within the parcel to saturation. At saturation the relative humidity is 100 percent. Usually a little more humidification is required to bring the relative humidity above 100 percent, a state known as supersaturation, before a cloud forms.

When air becomes supersaturated, its water vapour condenses out.

If the quantity and composition of the dust content is ideal, condensation may begin at a relative humidity below 100 percent. If the air is very clean, it may take high levels of supersaturation to produce cloud droplets. But typically condensation begins at relative humidity a few tenths of a percent above saturation.

Condensation of water into condensation nuclei (or deposition of water vapour as ice on freezing nuclei) begins at a particular altitude known as the cloud base or lifting condensation level.

Water molecules attach to the particles form cloud droplets which have a radius of about 20 micro meters (0.02 mm) or less. The droplet volume is generally a million times greater than the typical condensation nuclei.

Clouds are composed of large numbers of cloud droplets or ice crystals or both. Because of their small size and relatively high air resistance, they can remain suspended in the air for a long time, particularly if they remain in ascending air currents. The average cloud droplet has a terminal fall velocity of 1.3 cm per second in relatively still air. To put this into perspective, the average cloud droplet falling from a typical low cloud base of 500 meters would take more than 10 hours to reach the ground.

Precipitation

We know that all clouds do not produce rain that strikes the ground. Some may produce rain or snow that evaporates before reaching the ground, and most clouds produce no precipitation at all. When rain falls, we know from measurements that the drops are larger than one milli meter. A raindrop of diameter 2 mm contains the water equivalent of a million cloud droplets (0.02 mm diameter).

To get some precipitation from a cloud, there must be additional process within the cloud to form raindrops from cloud droplets.

The next rung of the Precipitation Ladder is Buoyancy or Cloudiness which signifies that the cloud water content must increase before any precipitation occurs. This requires a continuation of the lifting process. It is assisted by the property of water of giving off heat when changing from vapour to liquid and solid states, the latent heats of condensation and of freezing, respectively.

If the vapour first changes to a liquid before freezing, then there is the latent heat of condensation released and followed by the release of the latent heat of freezing.

This additional heat release warms the air parcel and adds to the lifting effect.

In doing so, the buoyancy of the parcel relative to the surrounding air increases, and this contributes to the air to rise further.

Now in the cloud, there must be a Growth of cloud droplets to sizes that can fall to the ground as rain without evaporating.

Cloud droplets can grow to a larger size in three ways.

The first is by the continued condensation of water vapour into cloud droplets and thus increasing their size until they become droplets. While the first condensation of water onto condensation nuclei to form cloud droplets occurs rather quickly, continued growth of cloud droplets in this manner will proceed very slowly.

Second, growth by collision and coalescence of cloud droplets (the collision of rain drops with cloud droplets and other droplets) is a much quicker process. Turbulent currents in the clouds provide the first collisions between droplets. The combination forms a larger drop which can further collide with other droplets, thus growing rapidly in size. As the drops grow, their fall velocity also increases, and thus they can collide with slower falling droplets.

A 0.5 mm-radius drop falling at a rate of 4 m/s can quickly overtake a 0.05 mm (50 micro meter) drop falling at 0.27 m/s.

When drops are too large, however, their coalescence (collection) efficiency for the smallest drops and droplets is not as great as when the drops are smaller in size. Small droplets may bounce off or flow around much larger drops and therefore do not coalesce.

A drop about 60% smaller in diameter is most likely to be collected by a large drop.

Clouds with strong updraft areas have the best drop growth because the drops and droplets stay in the cloud longer and thus have many more collision opportunities.

Finally, it may seem odd, but the best conditions for drop growth occur when ice crystals are present in a cloud.

When small droplets form, liquid water should be cooled well below 0° C (32° F), the freezing point, for ice crystals to form.

In fact, under optimal conditions, a pure droplet may reach -40° C (or -104° F) before freezing.

Therefore, there are areas within a cloud were ice crystals and water droplets coexist.

The ideal condition necessary for precipitation, in other words, for rain, has been duly satisfied.

The technical terms that were associated with the passage, I obtained from a web page (which I have changed to make the flow of the passage, as I would have wished for a reader to comprehend) were enormous and even with a mastery in English language and grammar one may find it difficult to understand the scientific concepts in its entirety.

Put that into simple English was difficult enough but I have made, an attempt to simplify the discussion, without distorting the scientific connotation.

Summary

For the synthesis of the above discussion in simple terms simple enough for younger age group (instead of 7 stages) person to understand I would break it down the concept of cloud formation into 3 or 4 essential stages without upsetting the authors description of the events in a hypothetical environment.

1. Formation of water vapour in a focus of dust particle of a particular shape and size.

2. Ascent of that tiny water vapour mass enclosed around the dust particles with the change in wind currents.

3. Cooling and Condensation at high altitude.

4. Acquisition of a particular size when gravitational pull brings it down (down to earth altitude) to the earth surface.

Next part of the discussion I would make my observations on the basic scientific tenet of uncertainty principle, change, chaos, gravity, entropy and thermodynamics in a superficial and philosophical point of view.

Chapter 10

Teaching Science-03

Even though the discussion would end up with the illustration of the basic scientific tenet of uncertainty principle, the change and the chaos inherent in any physical or biological cycle, the conceptual development of the themes that are involved are giant steps for young minds.

The frog leap attempt is the way to reach the summit of discussion.

To come to the final conclusion one has to go through the layers of scientific concepts and behaviour pattern of nature.

The fact of the matter is constant change.

But because these changes have patterns that tend to recur in somewhat of a predictable pattern, the sequence of events that recur are called cycles and the term equilibrium is introduced to bring a concept of static nature to a dynamic change that would be chaotic if not for the patterns of regular behaviour.

The persistence of patterns gives order and consistency to recurrent events.

Weather is an ideal ground to base the conceptual thinking.

That was the reason for its choice in the first place.

Having said that the professional meteorologist's meager capacity to predict the weather and his negligible ability to control it warn against the complacency of the scientific notion.

The teacher's limitation (myself in this discussion included) is an additional irritant.

Equilibrium states

A man with science would like to simplify facts of life for the purpose of explanation.

The equilibrium states is a model well tried out for stating concepts.

Sometimes in this process of simplifying facts he destroys the true tenet of reality but that is beside the point at this stage of discussion. For the purpose of discussion we take few examples of equilibrium states.

Cycles

1. Day and Night
2. Diurnal rhythms
3. Circadian rhythms
4. Rain and the Monsoons
5. Populations and the Birth and Death of individuals
6. Photosynthesis and the Respiration
7. Clouds and the Rain
8. Wind and Pressure
9. High tide and Low tide
10. Chemical Equations

Concepts

1. Pressure

2. Temperature

3. Gravity

4. State of the Matter

5. Radiation

6. Energy

7. Thermodynamic Principles

8. Entropy

9. Atmosphere

10. Air (water vapour) and Matter (dust)

Only term not listed up there is the Time for a valid reason.

To explain the day and night we need a artificial measurement be that it may be seconds, minutes, hours or days, weeks and months and years.

The seconds (small but perceptible enough) and days have a meaning for the clock manufacture but the minutes and hours are arbitrary ingredients.

Similarly hours and days have a meaning to the teacher and the parents and the weeks and months have no meaning (to the child).

For the child it is the play and no play (work) or the schooling and holidays and the rest of the classifications (weeks and months) are arbitrary and meaningless approximations in his life's experiences.

In a scientific sense, what the child needs is to be exposed to enough experiences adequate for his perceptual (development) thinking capacity and certainly not a rigid science syllabus.

I have no intention of discussing each scientific principle except a few that would have some meaningful expression in the child's mind.

Exploration

The explanations of the themes can be attempted by studying a model or an exploration. A cloud chamber in the laboratory can be a model (difficult to simulate) but in real terms unnatural for a young mind.

I would go for an exploration (observation, adventure and discovery) instead of demonstration with a virtual class of students.

This exploration is laid in the in the neighbourhood of the Hantana Ranges.

Why the Hantana Range?

It is simple.

A living laboratory to investigate the laws of nature?

How the modern man is destroying it or trying to restore it to order is an object for further discussion.

Perspective

The knowledge is incremental and the wisdom is not. Acquisition of knowledge without the ability to shed prejudices (be that it may be colour, creed, class, sex or the knowledge itself) already ingrained in the conceptual development is the biggest hindrance to modern day learning.

If the correct conditions are provided the development of wisdom (peace associated with

it) will ensue and of course the harmony of the
world at large.

Experiment an Eco friendly one.

There can be many variations to this skeleton of experimental inquiry. This is something I would have wished as a child but never was given thought by the scientific teachers of the yesteryear.

But I made my inquiry not necessarily in a scientific sense but silently (rationally as was feasible at my age) in my own time and space.

There are many more like me in this generation who are left without a guide in the search for knowledge for fun and game.

The exploration should consist of small teams. Should be limited to 5 to 10 for easy administration and division of labour without duplication of efforts and the efficiency of management.

First of all they should obtain a contour map of Kandy city (I doubt any school in Kandy have it except some understanding of the perimeter of 2 mile radius).

Then the team should identify sites for exploration for either 250 or 500 feet (Kandy-

Peradeniya bridge as a base of 1500 to 1600 feet) elevation intervals (altitude).

They should camp in these sites initially on day time (cycle) and subsequently at night (parents would go berserk).

With the day time data and their imagination they should be able work out their predictions and strategies for night.

With the integration in mind, the Boy's Scouts and Girl's Guides in schools should have their own contribution to this thematic learning and experience.

I have few site for the students to investigate. They are as follows.

1. Hantana Peak (with Radio Ceylon organizing the infrastructure requirement).

2. Peradeniya River Basin (at the Campus, Peradeniya Garden and the Water intake- with Peradinya Garden, University and the Teaching Hospitals providing the infrastructure requirements).

3. Kandy Lake and the surrounds

4. And any other site within reach of 10 kilometers.

The equipments necessary are minimal.

What the meteorologists in Kandy, use.

For the data the students could not obtain, the meteorologists should provide inputs of basic data (especially nocturnal).

Thermometers, barometers, compass and litmus paper would suffice.

They should make some weather balloons and few kites themselves for fun and experiment.

The whole exercise may take months of planning and execution but at the end of the day the students are enriched with the meaning and the understanding of why we do something in real life or at work.

All the concepts mentioned above and many more that would arise from simple questioning and inquiry would cover the entire syllabus with minimum of theoretical knowledge.

Since the education department have canceled the advanced level practical examinations for many decades and the university has no intention of recommencing them, this is the only way to rekindle the scientific inquiry and true scientific breeds.

It is the need of the day, I believe.

I have avoided the exact methodology (things to be recorded by the students) since it may look like the passage, I have used for discussion.

However just to illustrate a point the student should gather at the bridge that lay across the court complex in Peradeniya and study the contents of the stream (The Kunu Ela).

It is an ideal place for a pilot project for the students before embarking on an ambitions project.

I have used this for a theoretical discussion in a global sense to make aware of the complexity involved in doing research and weather reporting.

It should not be taken in literal or metaphorical sense but with open mind to discover the uncertainty principle which is the binding rule of the Universe.

The words that I have encountered in the passage are given below for objective examination by the teacher who may think like me and the list is not comprehensive.

In every 20 words there was a technical term and it is staggeringly high for a small mind.

Would clutter even an adult mind.

If the conceptual understanding is introduced in a half baked manner the damage it does for conceptual thinking is irreversible.

List of words

1. Liquid

2. Solid

3. States

4. Vapour

5. Content (zero to about 4 percent)

6. Atmosphere

7. Air

8. Temperature

9. Dust

11. Condensation nuclei

12. Nuclei

13. Ingredients

14. Precipitation

15.The Precipitation Ladder (model)

16. Stages

17. Processes

18. Form

19. Ascent

20. Expansion

21. Cooling

22. Air parcel

23. Air pockets

24. Wind

25. Flowing

26. Horizontally.

27. Vertically

28. Weather processes

29. Currents

30. Reversing

31. Air Mass

32. Convergence

33. Convection

34. Frontal Lifting

35. Physical Lifting.

36. Convergence

37. Surface

38. Space

39. Equilibrium

40. Warmer

41. Dense

42. Ascending Air

43. Frontal Boundary

44. Physical Lifting

45. Orographic Lifting,

46. Topographical

47. Barrier

48. Volume – Quantity

49. Pressure to be in

50. Expands

51. Condensation

52. Relative humidity

53. Humidification

54. Saturation

55. Threshold,

56. Vapour content

57. Mixing of more humid air mass.

58. Boundary

59. Supersaturation

60. Freezing

61. Condensation of water onto condensation nuclei

62. Deposition of water vapour as ice on freezing nuclei

63. Altitude

64. Cloud base

65. Lifting condensation level.

66. Ice crystals

67. Average cloud droplet

108

68.Terminal fall

69. Velocity

70. Buoyancy

71. Property of water

72. Latent heats of condensation

73. Deposition

74. Cloudiness

75. Heat

76. Cumulus clouds

77. Vertical growth.

78. Continued condensation

79. Collision

80. Turbulence

81. Currents

82. Bounce off

83. Flow around

84. Updraft

85. Collision opportunities.

86. Conditions for drop growth

87. Optimal conditions

88. Coexist

Chapter 11

Open Source Science

An Evolution

Introduction

There is an explosion of information exchange but there is no equitableness in sharing that information. Additionally there is no verifiability. This was the impression I had before delving into gathering some information about the colour and its distribution in the animal kingdom (fish to be precise).

To my surprise there is an open source science project probably taking its roots from Linux Phenomenon. A group of people with scientific orientation joining up to form an open forum for greater good of the public (rather than a corporate agenda) is commendable.

With the stem cell research taking priority in a commercial environment this is a healthy phenomenon.

Background

There is a proliferation of newspapers and journals in the local market (Sri-Lanka) and I am not sure what has given rise to this phenomenon?

May be making a fast buck.

I believe editors of these journals are not concerned about whether people are actually reading them or not.

Is this just a bit of publicity stunt, a corporate agenda or a political inexpedience

Exploitation may be?

I wonder why even the BBC is thinking of new innovations.

With the Internet taking its shape in speed (with speed accuracy is sacrificed), content and opinion making, the paper industry is becoming a static phenomenon. Ball by ball commentary of cricket in the Internet is boring to follow but busy people are adapting to the new trends. The local editors are failing miserably to give a true picture at the ground level and the world at large.

It looks as most of the writers and journalists have a "cut and paste mentality" and some are copied from foreign agencies without any comments attached by editors or journalists.

I believe journalists like our own Carl Muller should do some research on these trends.

With so many mushrooming industries, which include illicit liquor, betting stalls and video parlors and many more other nefarious activities (which one should not be proud of) what is happening to print industry is something bewildering.

Expectations

My Primary Objective is to make learning science a pleasurable experience to young students and budding journalists.

The Secondary Objective is to find how quickly one can gather information from whatever the sources available in an academic environment.

A subsidiary objective was to put that information in simple terms so that the reader could understand (improve my own writing skills as a by product) the contents with ease.

Conceptual Impressions

Gathering reliable information is a piece of cake and I could accomplish this with a drop of a hat.

I am looking for a "hat collection" of information.

Approach

Internet, reputed scientific journals, library and my own collection of old books were utilized in that order.

Outcome

Scientific journals always wanted to check on one's subscription status and ones that gave some free access was limited by arbitrary time limits.

The search engines were very poor.

The value of particular information was never highlighted in a sequential manner or order of merits.

There is just the proliferation of knowledge but no particular break down or an order of importance.

Reviews were noted for their absence.

There was an exception to this phenomenon of knowledge explosion especially in the publications of the Public Library of Science (PLOS).

Other exception is some American Medical Journals who are opening up the knowledge base to all and sundries.

A healthy trend is there but what is lacking is commitment.

Verdict

Cut and tie mentality of a surgeon's mind is replaced by cut and paste mentality of the window's users.

Writing as a skillful job has deteriorated.

Spelling mistakes and content jargons and inaccurate information were abundant.

To my amazement in one piece of writing the blood of the species of fish was almost identical to human blood.

I believe we have hereditary linkage to bacteria and fish but not to that extent. How my fish would die with a simplest of change in the environment explains why their immune system is different from ours.

I have neither become wiser nor enriched but perplexed and confused.

Simplicity

Simplicity was not the art.

Terminologies were not defined or poorly explained (in manner simple enough) for a kid to grasp the meaning to life and beings.

I have no surprise why many kids look at science without enthusiasm.

Books

My old books that dated back to sixties and seventies were a treasure trove for my illumination and they still are.

Those writers had a penchant for science as well as writing.

It is bounden duty for the scientists to write in simple terms with their own comments included. They should write books which little children would love to read and should not become self centered like our journalists who publish papers at the drop of a hat for image building.

Scientists should not become image builders.

They should take a lead from the open source science projects (Public Library of Science) and make science a living experience.

None of my objectives were satisfied after a month of trying but this is not going to change for a very long time.

Perspectives

The curiosity of a kid of yesteryear who loved watching at the little mosquito fish (Guppy) dancing in little streams (that are polluted now beyond any recovery) and changing their colour and tail configurations from generation to generation, lead to learning biology (with interest) and later becoming a human biologist (a pathologist to be precise) and that diversion did not deter the interests in colour (pigmentation, is the term we use in medical science) from all kinds of life (nature at large) to photography to computer graphics.

In spite of Diversity of colour (fish, toads, birds and bees) and the Specificity of a given species, there is a common thread that links all species that extend down the millions of years of evolution which is a remarkable feat of Simplicity of biochemical compositions (melanin, guanine, xanthochromes and haemoglobin) that expresses in various combinations in all living cells to give a particular combination of colours.

In the process of diversion at certain points in earth's history, however the secrets of Unity in Purpose (i.e. the defense against cancer, withstanding oxidative stress, skillful communication of impending danger that include ecological strain and becoming a living biological barometer of environmental stress of pollution), the unique survival instinct of all living beings was not lost.

However, the Unity of Purpose is lost when it comes to man's world (of incessant exploitation) where skin colour can be a decisive discriminatory factor in making choices (may not be scientific by nature).

That is the final philosophical point which I wish to make.

Effective Communication

If the expectations laid down in any investigation or analysis (whether it is accomplished or not) is not communicated effectively to all those who have similar interests, the commitment that make one to get involved in an endeavour becomes void.

The energy and theenthusiasm dissipate.

In other words it becomes a failure.

What I call the Club mentality whether it is a Garden Club or a Bridge Club that motivates people with similar interests to interact effectively is lacking in the Scientific Community.

It is a sad phenomenon.

Guiding Principles

The guiding principle should be to benefit mankind and all species on board this planet earth and above all the wellbeing of the planet itself.

In this process exploration if the individuals who get involved reap benefits it is of course a bonus.

It is a feather on one's cap.

Chapter 12

Waterholes and Bird Watching

Waterholes and bird watching do not go hand in hand. I believe people frequent waterholes when they are fed up of watching birds. But, the connection I have made with waterholes and bird watching is of a different kind. There was unduly high activity of birds and bees (really there were no bees) in the neighbourhood, for me to venture into investigating this simple phenomenon was spontaneous. The reason of this high activity was the perennial (but infrequent) rain we experienced. There was no scarcity of water except at times when the Water Board decides to divert our water to Kandy Municipality and forgets to turn our pipeline on (to cater to the commercial needs disregarding our needs who live outside the municipality).

The birds that visited our neighbourhood included were Greater Coucal, Common Coucal (Crow Pheasant) and the Red-vented Bulbuls and the increase in number of crows (from two to four to be precise in two seasons).

The reason and the only reason the abundance of rain and water.

The birds won't survive without the unusually high insect activity and snail population (especially the Common Coucal) which was well supported by the rain. The butterflies and their numbers increased proportionately in the dry periods in between the rainy spells.

Why this unusual congregation of wild life our neighbourhood?

To my mind there were two reasons.

We happen to live on their migratory pathway was one reason.

The other was the little waterhole and the water purification unit that stands in front of our house.

Birds frequent this place because of the abundance of water certainly, not for food. I of course do not believe in feeding them because of the constant threat to birds from the domestic cats. It was on the other night we managed to rescue a King Fisher from a domestic cat. He had one piercing wound on the head and another through the wings and there was bleeding.

I did not know how to manage an injured bird.

Got my daughter to read something on first aid and advise me until I attended to TLC (tender loving care). Luckily we had a visitor who was a bird watcher and we decided to keep it till the morning expecting its demise. Come morning it was still living and quickly handed over the bird to the Veterinary Hospital for necessary care. Quickly had a chat with the Professor (who unlike medical doctors visits the hospital even on Sundays and Poya days) and she assured me that the bird would be released in a few days.

An unusual happening on Saturday was worth a little film (discovery). I was amazed to see a crow chasing an eagle three times as big as itself. The cunning eagle was using the air current to float and the crow behind vigorously chasing behind the eagle was a fascinating site worth a documentary. The eagle floated itself above the clouds and the crow gave up probably with exhaustion and lack of oxygen in the higher stratum.

In a minute about six to ten crows gathered around the sky and signaled the impending danger to the mother bird and the event ended peacefully.

The young Coucal looking for a partner was another incident in the same week. The cry was different and distinct from the adult cry. These birds know how to hide just the same way they hide their eggs on crow's nest. With some of my own tricks hiding behind the shade I could spot the bird form a distance (thank god my long distance vision is still intact).

Before I could approach closer quarters, the bird was away in a flash.

Perhaps he may have to wait for the next season for a mate. Crow population has not matured enough to support two young ones.

In need to add few more passages for completion.

1. Within a year the waterhole was filled with soil to park a car.

2. Three large trees where they perched before and after bathing are no more.

3. We have a mulberry tree for them to feed.

4. The vantage point, I was watching them has no value now without the three tall trees felled.

5. I still go looking for them, when ever I have time.

Now the roof garden (with little space and lot of plants) is where I watch them.

I finished watering the plants one day and was gazing, the sky looking for the infrequent visitor.

The eagle.

I want to catch a close photo but failed several times.

It has a peculiar cry.

I had seen it perching on a tall tree, before.

It has penetrating eyes and the moment it spots you, vanishes in a flash, no chance of a photo.

But on this day, he had perched on the tree and stationary for my amusement. To my surprise it was there facing me but did not vanish in a flash.

Then I saw two swifts one coming from behind and other in front but to a side in combat form.

The swift was only the tenth of the size of the eagle.

The one in front was a decoy.

The one from behind would come from a side and attacked the eagle with its long narrow beak.

This went on for few minutes but the eagle would duck or turn its head to avoid the full impact.

It was unnerved by my presence.

I moved little closer and it vanished not because of the attack by the swifts but because of me.

Later I realized it had come to prey on the young swifts in the nest.

Chapter 13

Cloud Watching-01

Is timely to write something on clouds and I have waited more than three months from my analysis of a web page that contained cloud formation that described up to seven steps.

I broke these steps down to 3 or 4 essential stages without upsetting the authors description of the events in a hypothetical environment.

The idea of mine was for a child of middle year school to understand the theoretical aspects before even looking at the sky.

They were

1. Formation of water vapour in a focus of dust particle of a particular shape and size.

2. Ascent of that tiny water vapour mass enclosed around the dust particles with the change in wind currents.

3. Cooling and Condensation at high altitude.

4. Acquisition of a particular size when gravitational pull brings it down (down to earth altitude) to the earth surface.

This part of the discussion would focus on the naming of clouds before any attempt at interpreting the meaning of the clouds as far as the weather is concerned.

Having got these two steps right one is left to study in depth the meaning of these clouds just like somebody tries to read one's horoscope.

I believe in watching the clouds and reading their meaning is much more interesting than reading one's horoscope depending on the distant planets.

Problems with science is that we go back to Latin words (traditionally) when naming things and objects like the medical people of the yesteryear.

But once one has got the meaning and the basic tenet of their usage, the confusion that is inherent with not being able to handle the

129

language whether it is Latin, English or Sinhala (or Science) can be overcome.

This attempt is to overcome the problem of language and not to dish out some godforsaken uttering.

Description of Cloud Types

The clouds are classified according to three or four criteria with only five Latin words.

They are

1. The density of the cloud

Thin (which I call the flame type) Stratus

Thick (which I call the cotton wool type) Cumulus

2. The height (altitude) at the base of the cloud.

High Level Clouds (above 20,000 feet – 6,000 meters) Cirrus (which I call the horsetail or the curly hair)

Middle Level Clouds (6,500 to 20,000 feet -2,000 to 6,000 meters) Alto-stratus

Low Level Clouds   (below 6,500 feet -2,000 meters).

This description is almost similar to the density of the cloud described as-Stratus and Cumulus.

3. Height of the Cloud Mass

(of heights in excess of 39,000 feet -12,000 meters)

4. Rain Clouds Nimbus

In actual fact there are only five Latin words and all other names are combination of these five words. The total combinations (2x3x2) is only 12 but in actual usage it is less than ten (10) and with a description whether it is a rain cloud or not.

What one should know is whether it will rain or not or how heavy is the rain and how long that it will last.

In simple terms they are rain clouds, thin flame like clouds and heavy wispy cotton wool like clouds and any combinations of these cloud patterns at high level, middle level or low level with an addition of tall, dark and heavy clouds.

The meaning of tall, heavy and dark clouds needs no elaboration other than impending heavy rain which any soul can understand.

It is the interpretation of the rest of the clouds that one should be keen on and watching clouds should be a pastime for the young ones and try to interpret them and make some meaning out of them.

Making sums without meaning may be boring but getting the the weather right in the mind of a young can be an interesting hobby.

Make him a good observer (looking at planets with telescope would have some meaning to him) adding some meaning to what he / she observes.

This should be something of a fundamental nature (learning) in an agriculturally based country.

Looking for sunny weather and one day cricket are not the only things that (matter) one should worry about.

In scientific terms one should know three basic measurements.

They are the temperature (thermometer), the pressure (barometer) and the humidity (hygrometer) and these were the three basic measurement the yesteryear scientists started measuring and interpreting them.

Why can't we do that in our classroom is beyond my understanding.

The altimeter is an instrument that measures altitude (is an indirect measure of barometric pressure changes with height) and this is something a mountain climber should carry in addition to the compass.

In drastic weather as is seen in the west one who goes on a mountain climb without the weather forecast is a sure recipe for disaster.

We in the tropics because of the mild nature of the weather nobody cares to look at weather with interest leave alone geology.

The atmosphere only covers about 10 miles (50,000 feet) and in this one cannot live above 10,000 to 12,000 feet without acclimatization.

This atmosphere also 78% Nitrogen, 21% Oxygen and only 1% water vapour (including the carbon dioxide).

And this one percent (1%) is the most vital ingredient which we neglect (whether it is Water or Carbon Dioxide) and the non scientist including the politicians have no understanding of their significance to the life on this planet.

This why I make a case of learning basic science as an essential tool in healthy living and healthy planet.

Something that everybody can do (with a digital camera) is to take as many photographs of the clouds as one can see and give them a graphic description either based on the scientific terminology or one's own imagination and the use of the ordinary language whether English or Sinhala.

I am sure one is not at loss of finding a descriptive word for every cloud one sees.

Instead of saying "in every cloud there is a silver lining" please look at the sky when one is bored.

By the way, watching at formation of clouds can be an ideal focus (Nimitta) for a meditator.

Common cloud classifications

Clouds are classified into a system that uses Latin words to describe the appearance of clouds as seen by an observer on the ground.

There are four principal components in the classification system (Ahrens,1994)

Latin Root Translation

Cumulus Heap fair weather cumulus

Stratus   Layer altostratus

Cirrus   Curl of hair cirrus

Nimbus   Rain cumulonimbus

Further classification identifies clouds by height of cloud base.

For example, cloud names containing the prefix "cirr-", as in cirrus clouds are located at high levels while cloud names with the prefix "alto-", as in altostratus, are found at middle levels.

This classification introduces several cloud groups.

The first three groups are identified based upon their height above the ground. The fourth group consists of vertically developed clouds, while the final group consists of a collection of miscellaneous cloud types.

Cloud Levels

High Level Clouds

High level clouds form above 20,000 feet (6,000 meters) and since the temperatures are so cold at such high elevations, these clouds are primarily composed of ice crystals. High level clouds are typically thin and white in appearance, but can appear in a magnificent array of colors when the sun is low on the horizon.

Mid-Level Clouds

The bases of mid-level clouds typically appear between 6,500 to 20,000 feet (2,000 to 6,000 meters). Because of their lower altitudes, they are composed primarily of water droplets, however, they can be composed of ice crystals when temperatures are cold enough.

Low-level Clouds

Low clouds are of mostly composed of water droplets since their bases generally lie below 6,500 feet (2,000 meters). However, when temperatures are cold enough, these clouds may also contain ice crystals and snow.

Vertically Developed Clouds

Probably the most familiar of the classified clouds is the cumulus cloud. Generated most commonly through either thermal convection or frontal lifting these clouds can grow to heights in excess of 39,000 feet (12,000 meters), releasing incredible amounts of energy through the condensation of water vapor within the cloud itself.

High-Level Clouds

Cloud types include;

Cirrus and Cirrostratus

Mid-Level Clouds

Cloud types include;

Altocumulus and Altostratus.

Low-Level Clouds

Cloud types include;

Nimbostratus and Stratocumulus

Clouds with Vertical Development

Cloud types include;

Culumulus and Cumulonimbus

Chapter 14

Cloud Watching-02

I must confess that my first introduction to weather was from my British Colleague Mike who taught me the basics of reading weather in England. He used to say clear skies and cold nights and overcast sky and warm nights. He also went on to illustrate the scientific basis for his statements. In addition he taught me how to keep warm in cold nights in winter (which I did not have a clue) wearing two stockings instead of one which I usually did. We were great friends and I made sure I cover his work on Mother's Day and Father's Day regularly to allow him to visit his parents in Northern England.

He taught me how to pick the correct coffee beans and ground them instead of buying the cheap Nescafe. Unfortunately we are presently drinking cheap Nescafe in Sri-Lanka and tea dust which is good enough for manure.

In fact I put all the left over tea leaves to my plant pots.

I also should confess that I had very poor science teachers. My first science teacher was reading from a book written by Prof. Adhikarum (who was a Pali Scholar) and I felt he was trying to learn science himself let alone teach us.

Subsequent teachers were no better.

How, I became an environmentalist is by my own making probably a genetic tendency which I inherited.

My teachers never aroused any interest in me in science nor in my fellow beings. They were teaching science part time until such time they found alternative jobs.

The teachers were paid a meager salary those days but that should not have been an excuse for not being responsible teachers.

If I did not read the Kalama Sutta, my belief and also my disbelief of most of my teachers in Science would not have been firmly grounded.

This trait of course, I carried to the university and late Professor Osmond and Professor Carlo with their rejuvenating pep talks reinforced inherent abilities and left an indelible mark in our making and thinking.

Their influence, made us independent, in the way we look at the problems at hand.

We could weather the period of indoctrination which is carried to the present day uninhibited in the universities.

Coming back to cloud watching, there is no meaning to cloud watching if one has no understanding of their behaviour which is a function of the wind and its velocity.

So to understand pressure gradients and the development of the high pressure and low pressure areas in the atmosphere one needs scientific outlook.

I had a very good understanding about the wind and hot air balloons (even before I went into the Science Prep Class in the city) simply because I enjoyed flying kites and making balloons (with the help of adults).

Twice a year we made kites and at least once a year we made a balloon and filled it not with industrial Hydrogen (finding gas canisters was impossible those days and for our school laboratory need we had produce bio-gas) but with hot air and send it up with an address of the sender with a note "Please return, to the sender" even though we were never interested in recovery but were inquisitive enough to see how far the balloon went.

When we recovered those balloons they were soaked with water. Once the flame was out hot air cooled at high altitude and we did not need a science teacher to explain that fact.

My father hated these exercises but we did all these out of his sight and he never knew there was a scientist in the making.

With the first rustling of the leaves (especially the coconut leaves) we knew that the season is on with high wind that supports the lift of the kite. The two monsoon winds that blew in opposite directions and the blue skies were our watchful vigil. Fortunately both seasons came on schools holidays and there was no disruption to our school activities (we did not call them studies and the school was a place for fun and game).

However, I was never a cloud watcher those days but a cloud hater. I only wanted blue skies and sunny days for our simple activities which included playing cricket in the paddy fields and on the main road.

Thank god, those days there were no traffic jams like today and the bus driver's son and conductor's son were playing with us. They made sure that bus never pulled until the over was finished (balled) and we never had accidents.

We were road wise and road worthy.

There was no one-way traffic as in Kandy today. Some oblique reference to the traffic arrangement is worth a comment. The day they started this change like a gut reaction, I told my wife when a few lives (including students) were sacrificed on the road especially to the (unprotected railway track) train they would revise this decision.

True to my prediction there was a death, run over by the train on the first day of the exercise. In this country there is callous disregard to injury and life in general.

I am a firm believer that more people die on the road than due to war or tsunamis. Even today near Gatambe young boys were attempting to fly a kite on the main railway track and few minutes later a fast train was on the track. The railway department should make an attempt to educate the people (who come to the new court complex) to avoid crossing the track and funnily enough there are concrete steps directly leading to the track. The 5 kilometer track of railway should be adequately protected with the new one way system in operation before few more lives are lost.

Pressure Changes and the force of Wind

Air pressure changes with altitude and at 18,000 feet the air pressure is half the pressure at sea level. Not only pressure changes with altitude but it also changes with distance as one travels horizontally assuming one is at the same elevation. This change is due to changes in temperature and humidity of the air.

Unequal warming of the air by the sun brings about this pressure changes. The earth's tilt on its own axis make a drastic difference in the amount of sunlight reaching the surface. Not only this makes seasons in the Northern and Southern hemispheres but it also makes warming of the air masses above the hemispheres different.

Land and sea and the day and night cycle also have differential effect on warming and cooling and this in fact, creates low pressure and high pressure regions in different parts of the earth's atmosphere.

The air blows from high pressure area to a low pressure area creating wind. The spin of the earth and centrifugal forces also have some modifying effect on these air currents.

Gravitational pull effects the wind but its effect is uniform all over.

When the mass of air moves from one place to another, it carries the clouds with them. Some of these masses are warm and some of these masses are cold. A cold front meets a warm air front it creates various changes in the atmosphere which include thunderstorms.

The other factors that affect the wind and it direction of motion are the contours of the earth surface and the tree cover.

Some trees are excellent in breaking wind.

Just in this year in our neighbourhood two trees were felled for expansion of the neighbour's house. Now the wind blows directly at our veranda and I had to move all our plant pots to a new location to avoid falling downstairs without a wind shield.

Trees have a moderating effect on wind and careless uprooting of trees increases the damage caused by wind.

When we were flying kites as young ones we knew moment we got the kite above the coconut trees we are set for the higher wind and there after it was smooth sailing, rather smooth flying of the kite.

Of course, we did not understand (as young kids) the friction that the trees created but always climbed to higher elevations including rooftops when not successful in getting the kite up.

When one listens to the weather reports, on radio or TV, everybody including young ones feel it as a load of rubbish but this has to change.

Weather reporting is a specialty leaving it to be edited by a news editor who has competence only in English (any other language) but no scientific sense is a bizarre experiment by itself.

In other countries there are weather channels. Why our meteorology department take a plunge to this reporting activity on air and TV is an open question. I believe there is some lack of coordination between two departments.

Trying to discuss the finer details (in a short essay-need diagrams and other aids) is not intended but this introduction would make somebody with scientific inclination to investigate and dig deeper in and further.

If one is studying clouds, one need to understand the temperature, relative humidity, pressure changes and the measure of the wind speed.

When one looks at the different types and pattern of clouds with armed with these basic (parameters) measurements one gradually becomes a forecaster automatically although amateur.

These are the first steps for a person who wants to become a meteorologist.

Chapter 15

Education

Is neither an exercise

That is

Hands off

Or Minds off

Of children

Nor it is a delivery

Of a curriculum

With rigidity and red tape

Communicated by only a sage

It is an activity

To help

The mind of a child

To think

To inquire

To explore

With open mind

And then to exchange

Elaborate

Expand

His own ideas with meaning

Chapter16

Thinking

In ordinary sense
It is the spontaneous activity
Of engaging in one's own
Beliefs at random
Or beliefs burrowed
From friends, parents, teachers
Or others

But thinking
Without reasoning
Inquiry,
Or refinement of one's prejudices
Is like visiting
The Devil's workshop
In one's prime holiday
On a poor budget

Chapter 17

Chaos, Mind, Matter and Life

It is easy not difficult to understand life in general (plant and animals) as the only form of energy in this world that has an organized form of arrangement with arising, peaking to a stable state and gradual decay. This cycle of origin, growth and decay repeat itself with intervals of regeneration (reproduction to be precise) of an almost identical image of the predecessor giving rise to a new equilibrium.

But in physical world there is no reproduction or replenishment of energy but a constant state of chaotic energy transformation from one form to another in a perceptible but in a disorganized fashion.

For example rain is known for its constant change with time, location of a particular point on the planet (its changing orbit around the sun making subtle changes) and the altitude.

The changing pattern from one physical form to (water vapour to water) another determined by temperature, wind and contours of the planet, in real term is chaotic. If one imagines the surface land as a concrete block without plants and vegetation the pattern of air currents over the land is going to be entirely different and would depend mainly on physical factors.

If one takes the entire globe as a unit and assumes that there is no vegetation covering the surface on earth the rain pattern would be determined by random and physical events determined by temperatures and wind patterns (cyclones and anticyclones).

But the trees and vegetations are a collective physiological process both at macroscopic and microscopic levels which work in unison to produce climatic changes, which cannot be understood only by physical (physics) processes.

One has to be a biologist to understand the intricacies associated with such a mass of physiological processes. The forest cover affects the cooling process of rising air and it also contributes a body of water vapour to the air that ascends over the mountains.

The jungle is not something dead but it is a living creature contributing to the atmosphere. With vast vegetation covering the planet earth with uneven contours and distortions the weather pattern are perceptible to be predictable in some way. The physiology of water metabolism in plants which in turn reacts with the physical factors gives some uniformity to the planet that would be otherwise chaotic and without which life cannot be sustained as it is of now.

This is why man should not upset its balance by man made misadventures. Global warming is one of them we seem to barely understand but not come to term with. Inter monsoonal rain is an obvious example where would be chaotic rain patterns are organized into a somewhat recognizable weather pattern. So the biosphere is vital in our stability and any disturbance in this biosphere by physical means (whether nuclear bomb or a terrorist blast in a nuclear plant) is going to be catastrophic.

But if one looks at the behaviour of water in purely physical terms the lower the temperature the less is the chaotic behaviour. Ice is organized into crystal pattern, the water is semi organized in physical form which takes the shape of its containers contour but the water vapour behaves in random and chaotic manner. But it is this chaotic behaviour that makes life possible and the water available all over the planet earth, even though, unequal in distribution.

Moments the ice caps melt and there is no ice on this planet the catastrophic changes going to be unimaginable.

That is point I want to arrive at but getting there needs understanding of matter and energy. Matter and energy has linear relationship with motion and at a particular speed of motion (the speed of light) the distinction between matter and energy becomes indistinct. At this point the behaviour of matter can be understood only by the principle of chaos. A linear equation has its limits and another dimension of mathematics takes precedence in such a scenario. Either there going to be random occurrence of events or haphazard chaotic behaviour that would set in at the speed of light and it is the most natural thing to imagine or guess.

But in real life, there is neither complete disorderliness (randomness) nor complete orderliness but it takes its shape somewhere in between.

There is some orderliness in the disorder phenomena. This orderliness is mainly due to the biosphere which we are enriched with.

This is a fact which we tend to forget covertly or overtly. Because of the magnitude and the diversity of the biosphere and the ability of the biosphere to sustain some equilibrium and orderliness we tend to think that the biosphere is dispensable.

This is the basic error we are making in energy planning. There is obsession on production of energy and mechanization but one does not realize the energy efficiency and transformation cannot go beyond 30 to 40% and the rest (60%) is wasted in the process of transformation. It is a colossal waste and that is why the question of sustainability comes into operation.

For a start only 3 to 5 % of the solar energy is converted to biochemical energy in photosynthesis and this transformation takes in small doses at an electronic level by the changing orbit of electrons from one orbit to another (electrons are in constant state of chaotic motion determined by atomic numbers). Even though the energy transformations in biochemical pathways are slow and low, the energy efficiency in atomic level is very high. This is why bio-transformation does not upset the equilibrium that exists at a particular time but any changes that may occur are spread over millions of years.

But in scientific terms the nuclear fission (man made) or nuclear blast is an obvious aberration to even the so called chaotic behaviour of atoms in general.

I would like to divert to another topic which is topical in some sense as far as tsunami is concerned. We have had serious of underground atomic explosions (so called testing) in Asiatic plate (India, Pakistan, China and Russia) and southern hemisphere (France) and in North America. So far we have not had any scientific estimate of the stresses that they may have caused on the tectonic plates. These stresses however small to begin with are cumulative and may initiate random disorder (chaos) on the movement of tectonic plates and the recent earth quakes and tsunami (of the order of 8 Richter scale) may have resulted from these small aberrations.

I believe that the scientists are covering up these issues.

It took some decades to understand the effect of acid rain on inland water reservoirs but when it was recognized, it was too late and, an irreversible damage was done, to all the countries in Europe.

Now it is evident by the understanding of eco-principles in the West but lack of it in the East.

Whether one likes it or not with the explosion of population the eco-crisis is looming. It is not only energy crisis we are going to face we are going to face many other, that include the food crisis. What ever energy policy one may espouse if the eco-crisis is not taken into account the domino effect is going to be chaotic.

The relationship of temperature to global warming even one ignores the carbon dioxide factor has brought about chaotic weather patterns, floods, cyclones, hurricanes and the like.

They are eye openers to more chaos to come from time to time.

Man's overriding desire for 100% energy sufficiency is only a dream and would not be ever achieved unless sustainability factor is taken into account. For sustainability the renewable energy source is the only viable option and any other approach would be disastrous. For an example going for nuclear energy without adequate safeguards for 100% safety and security, it can lead to disaster.

One suicide bomber may nullify years of safety and preparation in a moment of insanity.

In the long run any energy generation by physical means (atomic) is going to be destructive and a big strain on the biosphere.

It would bring about chaos unforeseen.

The destruction the mind of a maladjusted person can do is enormous. This brings me to the point and mind and its chaotic beahviour.

Mind is known for its chaotic behaviour.

If one lists down all the thoughts for only five minutes one would realize how chaotic the mind is. Further elaboration is unnecessary. But the beauty is that mind inhabits a biological system which is very slow in its perception whether visual, sensory, and auditory or any other sense.

This puts the chaotic behaviour of the mind under check.

The living beings activities are under ordinary circumstances are fortunately determined by primary senses and the mind's contribution is generally secondary.

A Buddhist thinker is free to disagree with me ("Chetanaham Bikkave Kammam Vadami") but I am referring to a basic biological plan and not talking about the philosophical plane.

Mind's basic behaviour is chaotic and what I believe is that all the energy planners in the East and the West are behaving in a chaotic fashion but without a global vision.

Chapter 18

Theory

Is a mental manifestation

Of an experience,

An impression

Or an inspiration

That has been tested and verified

By many with similar interests

And in concurrence

But the problem with theory

Is that when it get established and ingrained

As reality

With rigid laws and concepts of convenience

In the people's minds with dissimilar intents

Validity of the original tenet

And its application

Become distorted and obscured

Out of context

Chapter 19

Hypothesis

It is a tentative impression

One creates in the mind

Of random thoughts

Of some observation

Or an experience

And its occurrence has

No control, relevance

Or any concurrence

In its natural history

And its evolution

At its best

Is a guesswork

Of some kind

So the man has coined

An inspirational word called

Intuition

For its sustenance and nurture

In the history of mankind

And it is neither tested nor verified

As at present

Or has any application
Of significance known yet
But has arisen in the mind
Of a clever observer
As a new proposal
For testing and validation
In its original context

For example man has created
Unique races with unique aspirations
Far above the ordinary meaning
In life and matters

And when this fallacy is self perpetuated
At will
It creates disorder, disunity
And discordance of various magnitudes
Leading to wars of attrition
But in reality what is perceived as unique
Is a manifestation of cyclic nature
Of change
Of the world with diversity
Of opinions and prejudices

Chapter 20

Concepts

Concepts are as good as
The words that describe them
And when the words that describe them are
Simple enough
The clarity and content are obvious delight
But when the dialect used
Is not precise,
It becomes a pain in the neck
And the meaning of the concept get muddled
And fall into disrepute an ambiguity

The laws that govern the concept
Are the highly structured
Grammar of Science

Chapter 21

Explore

"Curiosity kills the cat"

Is the proverbial stunt

That is

Subtly insinuated in the mind

Of an explorer

By the one who believes

That the authority

Of the scriptures are sacrosanct

But anything that is not explored

Visibly and in depth

Is not worth

The substance

That is said to be

Standing

On its own accord

Having said that,

Exploring without the tools

Of refinement

But with the tools of tradition

Is fraught with danger

Inherent there within them

Chapter 22

Inquiry

Is the ability
To ask the right question
At the right time
At the right place
From a right person
And not expecting
The answer
To be all closed
For any further questioning
Or inquiry

Chapter 23

Tools of Investigation

What then are the tools
Of investigation?

They are the ability
To ask the right question
And the ability to scrutinize
The answers obtained
With laser point accuracy
But leave the final outcome
Open ended

Chapter 24

Inference

Inference that is achieved

Without inducing

A method of free inquiry

Is not worth the substance

It claimed to have been deducted

As final, long lasting

And true in nature

Chapter 25

Knowledge

In general
It is the habit of collecting
All redundant facts and figures
Of insignificance

The real knowledge
Is the ability to
Drop old ideas and beliefs
From the basket of knowledge
In possession

But the true knowledge
Is to know that
The latest acquired
Knowledge is the one
That has to be dropped first
Before going any further

Why one should do that
From birth to death

I wonder

For a wisdom tooth

Chapter 26

Energy

Come to think about it
Energy is the least understood
But easily
And often
Misquoted out of context
In science and in journalism
And if I ask the question
Does the matter has energy?
The answer is not clear cut

Then I ponder for a little while
And ask again
Which has more energy
Matter in motion
Or the matter remaining still?
I am called a trouble maker
In my class of adults

When I delve in
Bit further

And ask

What is the smallest particle having energy?

Is it the electron, the neutron, the proton or the
quarks?

Then when I ask what is the smallest unit of
energy?

A photon

Or a wave

I am very much

Stuck with words of expression

Now that I am in deep waters

Not understanding

All these concepts

What shall I do

To move any further?

State of the matter

Is that

If each particle

Carries loads of energy

What is the essential force that

Intrinsically bind them together
And hold them tight?
In relation to position
They keep
In space and time
While allowing free
Moving particles to move
Without much of a hindrance
Or rather than with
Minimum of interference
From its neighbouring
Particles of existence

If only the moving particles have
Mass, velocity and energy
That goes with the momentum of forces
What is the principle that
I should apply to my body of interest
Where me standing still
On this planet earth
Has no energy
Until I decide to kick a football
With a force

Now who has the energy my leg muscles
Or the football or both?
How confusing are those concepts
Energy and Forces in action?
I wonder

Then again I should have
Some energy to stand on Earth
Much less if I stand on Moon
What about moon walking,
Sweet talking and sleeping on the go
In this universal universe and space?

What are the energy consumptions of all these
activities on earth, on its outer boundaries and in
the space in between?

The mind
That made me to kick the ball
Should have either some energy or force
In any case
I argue
Since the Mind

That made me to kick the ball
In the first place
Made the movement of the ball
Possible
Until such time the
Mother earth had decided to
Abort it abruptly in its journey
In space and time
So the mother earth should
Have some control over my mind

In fact the matter over my mind
Rather than mind over matter
In this instance

Bees, flies, mosquitoes and birds can fly
And in their case
They decide to fly away
As they wish
And also control
The start and stop
Of their flight
As desirable

So birds and bees
Have more energy than me
And humanoids

Staggeringly confusing
Isn't it?

So who has more energy
The Mind (Kicking)
The Matter (Ball)
The Motion (Transition)
The Bees (in Flight)
Or the mother Earth (Gravity)
That stopped the ball in its trajectory?

I wish the answer
Is clear enough
For a kiddie

I am not kidding

The answer to those questions
Are open to inquiry in a

Philosophical Sense

Now that I am beginning to

Think as a Thinker

But not as a Scientist

Who would provide me

The answers to the scientific tenet

That were raised in an open forum

Of inquiry

What is more powerful

Energy wise

The Thinking Mind

Or the Science in Action

I believe the former rather than the latter

Since the mind can penetrate

These concepts at will

Or at leisure

What is an energy and what is a force

The mind's energy by its very nature

Is probably far more superior

In its contents and activity

When used wisely and appropriately

Subtle and pervasive

Penetrating

The truths of all kinds

It is the mother superior

Of wisdom

Now winding up my thought process

Who had used more force

The Italian Parvo Rotti's Goal Header

The French Zidane Zidane's Head butt

And an ordinary Sri-Lankan's (my) Penalty
Kick

At the goal post

Of wisdom

Sans Destruction

The power of the Mind

Could be ascertained

By intense focus

And attention

To its moment of Origin

And the Energy of its consumption

Can only be harnessed

By sustained
Commitment and Conformity to details of
Metta, Anapana Sati or any object of
Attention sans attachment
And the Competence in Meditation
Cultivated

By that intention
It is achieved
By the wise
The Arhaths
The Buddhas
And Passeka Buddhas
Of today, tomorrow, yesterday
And now in this moment of
Universal extension
Of the timeless dimensions
Of the Sansara Cycle

Chapter 27

Power

Few hand flexes
At the wrist joint
That generates electricity
By pumping action of the hand muscles
From mechanical power
To light a Wrist Torch
Is a modern manifestation
Of conservation in action

Yet the efficiency
Of the conversion of
Biological energy
To electrical energy
Is only forty percent
And is considered to be
The most efficient
And produces least amount
Of Carbon dioxide

But

The energy efficiency

Of the Coal Power

Is only ten percent or so

Considering the inherent wastage

Encountered

In developing

The infrastructure

And the corruption envisaged

That is rampant in

Energy Economics

Yet

The production of Carbon Dioxide

And all the noxious gases

Are staggeringly high

And the heaps of

Rubbish ash left behind

Is monumental

Power corrupts

Energy consumes

As soon as

It is produced

Absolute Energy

Of hundred percent sufficiency

Set in motion

A chain reaction

So devastating in magnitude

It would take billions

Of years for recovery

That is the Vision

The consumerist society lacks

And fail to grasp

In this moment of selfish needs

This is why I say

That the scientists don't have

Philosophy

And Philosophers don't have

Neither Mind Power

Nor Muscle Power

In this contemporary world

Of macroeconomics

To say that

Energy sufficiency of 65%

Is the most that

This tiny island can tolerate

Without

Upsetting the balance of nature

And the rest 35%

Should be left in wilderness

Chapter 28

Wisdom

The acquisition of knowledge
Without shedding the prejudices ingrained
Be that it may be
Colour, creed, class, sex
Or an image
Or the knowledge itself
One has built oneself
For self perpetuation
Is insensible
And insensitive
To the needs
Of the majority

The knowledge is incremental
But the wisdom is not
It can only be reached
By sustained attempt
At ridding of prejudices
Of all kinds
Inbuilt in the

Cycle of uncertainty
In life and matter

Chapter 29

Freeman

Let my mind
Toy and joy with
With compassion, kindness and equanimity
Let it be that nobody
Believe in something
Since it is
The tradition to do so
Or it is in the scriptures
Or it is hearsay

Hearsay
And my teacher said so
Let there be no bias
Towards a notion
Specious reasoning
Faith
Liking
Soothing to the ear
For repeated hearing
Do not go upon

By rumour

Nor upon surmise

Nor upon a lovely axiom

Or just by pondering over

But by ones own realization

Greed, hate and ignorance

Are not the root causes?

The birth, decay and death

Are the miseries

And until one knows

By oneself

What must be done?

Has been done

There is no more to be done

No more to come

The life has been exhausted.

Truly

The freedom is achieved

From wit

Chapter 30

Tastes and Interests

There was a TV program called "Rasa Sarana". It literally means 'Enjoying Tasty Food' but I translate it differently as 'Good Taste' for the sake of Science and Food.

"Rasa Sarana" was an ideal program to watch for both science teachers and primary school teachers.

I watched it occasionally but a recent program on making beetroot cake took my interest on two counts.

Reason one was I am one who would campaign against adding artificial colours to food (from chillies to cordials to jams) items and the reason two was my natural experimental instinct (to see what happens to red colour of beet root after cooking; similar to the child's inquisitiveness).

However, I am making an attempt to illustrate two thinking processes that adult use as a matter of fact which children lack in their formative years. Teacher should be mindful to these aspects of children's mental activity when teaching.

The theme for the exercise that follows is based on a model "the cake" and "the recipe", the building materials. To appreciate the transformation of matter in cooking or baking (which involve heating) adults have to move from "concrete operational" thinking (direct observation and experience) to "formal operational" thinking. Both forms of thinking are operational (not mutually exclusive) in daily activities but generally speaking nobody makes an attempt to isolate them. Thinking is not an isolated process but a constant accompaniment of all our activities.

I had a formal analysis of the colour change of the cake before (preparation of the recipe) and after baking. My analysis goes on like this. The red colour of beetroot is water soluble and would seep to starch in the process of heating (cooking) assuming that there is no disruption to the pigment molecule in the process of heating.

As for my experience with the beetroot curry the possibility of breaking down of the pigment is most unlikely.

Pigment molecules are by nature resistant to alteration. The yellow colour of butter, which is fat-soluble, would not contribute much to changing the red colour.

My final anticipated colour (based on my previous experience of butter cakes and the objective formal analysis of the current activity of baking a beetroot cake) of the baked cake would be yellow in colour but specks of pink around the pieces of beetroot.

Even though, I did not have a very close look at the cake (with the TV resolution poor the distinction was not possible) I was proven wrong (end result of the experiment).

I had to wait for one hour but I was not bored. I considered both chefs as well as the two food scientists were excellent teachers.

The way the program was conducted was excellent. With little diversions here and there for listener questions but without straying away from the main object.

At the end I was much enriched with thoughts that made me to compile this and extend its application to children.

The doctor made the point of prudence of using natural colours. While I reiterate the same point here, the use of natural colours was good departure from routine (adding of food colouring) practice.

One also should not lose sight of wholesomeness of food we eat.

For example pol sambol is a wholesome food, which is better than a piece of cake. It has roughage, vitamin E and B and easily digestible monoglycerides.

Saffron added to dhal a pretty good antiseptic.

Practice of Teaching

Colour Sense

This program illustrated many features that are relevant to teaching young children. Children love activity and enjoy colours. The chefs used only one colour and it made the experiment simple and exciting. The colour of the butter was simply ignored. That made the learning process easy for everybody, adults as well as children.

I am one who believe in using only 3 or 4 colours in academic activities.

Many colours make any intended distinction and attention muddled in children and adults. If one look at primary schools the children are given at least 12 colours and sometimes 24.

I also made this mistake as a parent.

I realized the use of many colours is counterproductive.

For a starter think about of using many colours in a office set up; how attractive it would be to the manager and how confusing it would be for the apprentice in training.

It is a ridiculous adult concept to use many colours when a few colours would suffice.

Weddings and wedding dresses are good examples.

There is nothing wrong in using as many colours but teacher should not be choosy.

Let the children choose 3 or 4 colours and let them use them consistently.

Let them mix the colours and experiment with the colours. Instead of giving them set of 24 coloured pencils give them only a few and only one at a time (when 3 to 5 years old).

Christmas coming and the new school season around the corner, parents should be mindful not to kill tastes and interests of young children.

The teacher might find only a few would use many colours and strikingly, only a few use only one colour.

In that situation it may be desirable to investigate to see whether the child is colour blind.

If not leave him alone.

The teacher would realize how unique each child is, at least in one respect, the choice of colours. Just like we adults are choosy in deciding the colours of our dresses children too are choosy.

Attention Span

Whether adults or children our attention span is limited. For a single subject a child's attention, at the most is 30 minutes (probably 45 minutes if the lesson happen to be the first).

So keep the lesson short.

When an activity is involved one hour is the most for a child.

204

Anticipation

The activity should have a final outcome Expected result should be theoretically discussed briefly in the first few minutes (anticipation - like me waiting to see the final color of the cake).

Arouse anticipation with proactive questions with variety of answers from children. But keep it simple.

Keep It Simple and Short (Stupid) Principle.

Repetition

Repeat the exercise if children enjoy it with a little variation. Add some spice to the teaching and learning experience.

Story

Build a story around the exercise and have few cracks and jokes. Stories make the children with short memory retain the learning objectives.

That was the only part missing in the TV program.

If the chefs related where the idea originated and some of their hilarious mistakes in the kitchen I would have give them 10 out of 10.

Finally, always strive for perfection and quality even with children but not with obsession.

Preach what you practice.

Chemistry of Food

The same experience could be extended for older children to make some progress in science.

This is where I got myself stuck for three months and without any worthwhile progress.

In the first instance, I did not know where to look for and I did not have any resources at hand.

Biochemistry books were of no use and asking my students a few bizarre questions did not yield the desired results. They knew all the biochemical components and reactions and they did not know what would happen to rice grains when boiled at a particular temperature or at a particular rate of increase in temperature for a particular length of time.

They are not trained to become cooks and they are not even taught physics nowadays in schools and properties of matter are alien to them.

Sols, gels, colloids and viscosity and gelatinization are not of any interest to them.

Finally, I stumbled upon a scrappy paper and following is just a glimpse of physical science of food.

For a starter there is lack of scientific data related to food preparation and big companies who make potato chips to gravies to sauces to soups and keep them as trade secrets and do not elaborate on what modifications and additional materials are added to them.

They are not natural or wholesome foods to begin with. They are more worried about keeping quality and effects of freeze-thaw cycles and not the nutritional aspects.

Starch such an important constituent in many food preparations to arouse the interest in young school scientists, I would pen a few distractions to simple biochemistry.

It is interesting to note that many school children in England do not know how potato chips are made of?

Hope somebody with interest in food science would write few articles.

Basically starch is made of two major polymers (giant molecules with few basic building blocks like in polythene) made of single basic units called glucose units (not sugar, sugar is made of one unit of glucose and one unit of fructose).

Animals store glucose in the form of glycogen. The two major components are Amylose (without branches like a set of compartments of a long train) and Amyopectin (like the railway tracks with many branching points).

They are stored in ovoid or spherical granules and in the plant material is in desiccated (with little amount of water) form to improve the keeping quality (prevent undesired chemical reactions) and for the availability for future use.

All animals including humans use glycogen when starving or in long physical activities, like marathon running.

Humans of course modify this by milling and cooking.

One of the basic methods is to add water and heat to make a paste (dough) and make molds and reheat or re-cook to make different food items.

In a recipe of course one adds salt, sugar, milk, chocolate, butter, eggs and orange or lemon juice.

The most important ingredient undoubtedly is the water one adds to the starch.

The modifier is the heat and the rate of heating and cooling.

The amount of water added would determine whether we have a soup, a gravy or a cake.

The heat breaks down the polymers to smaller polymers and on cooling these smaller polymers tend to realign themselves to different forms and textures.

This is the art of cooking and the other ingredients added modify this texture in innumerable ways and we call them the delicacies.

The items added to the recipe (except colouring agents) determine the final outcome depending on whether the item is water soluble or fat soluble.

For example the sugar which absorbs water in preference to the starch would affect the gelling property of starch.

Salts do likewise albeit with a difference, the smaller components (polymers) formed with heating are ionized and thereby prevent their realignment when cooled.

Lemon Juice by changing the PH of the medium has other properties apart from changing the taste.

Interestingly fat (frying) in the food make ingredients dissolve in cells easily (taste buds) and make them very appetizing.

The food industry knows this well and induces young children to go for fatty foods (starting a very bad habit early in life, a precursor of atherosclerosis).

Where the "pol sambol"(monoglyrides) and "aggala" (energy in compact form) do the needful the industry want the children to go for highly modified margarines with a pinch of vitamins.

Children should be made to experiment (with adult supervision) with dough, paste and gels and unfortunately food science (cooking to be exact) is not an important subject nowadays.

Coming to scientific terms we are talking about viscosity, gelatinization, colloids and the like.

If one look at the end point of food granules under the scanning electron microscope (SEM) would be as fascinating as eating them.

To arouse curiosities of the young (the very young) I should wind by giving some insight to the starch in rice.

Rice has some properties, which is different from wheat flour.

Rice does not have the protein called the gluten that gives the peculiar quality to the dough.

Additionally its composition (percentage) of amylopectin is high.

Apart from its ability to absorb water it is easily digestible and Kiribath is a wholesome food ideal for children (without the gluten allergy).

What I was expecting from the previous regime and the aging Minister of Technology was did in their research work over the past ten years found the correct composition of rice and flour mixture for our baking industry?

Are they still continuing on the research?

That is worth pondering.

They did not.

Chapter 31

Cycle of Life

With the success of in vitro fertilization over the past quarter century, it is worth looking at future trends with past experiences in a philosophical point of view. During the same period and perhaps little longer the cell cycle events reveals, somewhat of a different outlook in biology and science.

Trying to recreate life in close compartments is fraught with inherent dangers. The cell cycle events gives us a preview of these dangers and what I state should not be taken into account neither in literal sense nor scientific sense.

It should be looked at in a broader view and outlook.

It is necessary to briefly state that every form of life has its own right to live in this planet and the man's existence in this planet is not a right by itself but one of privilege (not exclusive) inclusive of all other beings including plants.

The guiding principle is Avihinsa but not to overrule.

Saying in another words;

'Live and let live.'

Humans have no right to kill other animals unless its own survival is threatened by natural events and not by man made artificial events.

In a philosophical sense man is threatened by its own species more than any other species on this planet and our own history is annotated with sufficient examples unless of course a deadly virus that originates in a cell culture medium or in a breeding facility for birds, takes precedence.

Cells have a finite cell cycle and the moment this programmed cell killing is interrupted bizarre forms originate. The best example is the cancer and fortunately cancer kills its owner prematurely and holds up the unwanted biological event.

These events go in tandem with the random mutations at gametogenesis (formation of male and female sex cells) with or without biological advantage.

Humans have a finite number of 50 (roughly) cell cycles and 30 of these cell cycles are finished in about 12 weeks (by the end of the of embryogenesis). Another 7 to 8 cycles are completed in roughly 28 day cycles by 40 weeks of gestation (Total of 38 cycles) when a fully mature baby is is born.

From gametogenesis another 15 cell cycles are completed when adulthood is attained in humans.

Of these 15 cycles probably 8 cycles occur before birth in roughly in 28 day (Lunar) cycles and another 4 cycles in the first year of life in roughly 60 to 90 day cycles.

In effect 45 cell cycles are completed in total up to the time of maturity of the individual around 20 years of age.

From then onwards the last five cycles to old age are spread over span of 10 to 12 year periods. This lengthening in between cycles is important lest the aging process would be as fast as during the growth period.

There is a condition called progeria where the aging cycle is accelerated and the person becomes old in about 8 years and die prematurely.

What it means is that the man is left with only five (5) more cell (life) cycles to live with which is accommodated within 60 years (assuming maturity is achieved at 20 and he lives for 80 years). When the number of years of adult life span left, is divided by the number of cell cycles not completed then there are five more life cycles (for each cell cycle) of 12 year duration left for living.

Stage I adult hood (Juveniles) 21 to 32 years.

Stage II adult hood (Late Adulthood) 33 to 44 years.

Stage III adult hood (Onset of Senility) 45 to 56 years.

Stage IV adult hood (Early Senility) 57 to 68 years.

Stage V adult hood (Late Senility) 69 to 80 years.

Whether one likes to believe it or not, the fact of life is that we start aging within one cell cycle after achieving adulthood and it is unfortunately around 30 to 45 years of age.

This aging process is well studied in the brain and according to some estimates about 10,000 neurons in the brain die every day from the age of 35 years. Brain cannot increase its cell numbers (permanent cells) after maturity (astroglial cells can divide but they are only supporting cells of the brain) and that is why for study of aging process the brain is a fashionable and favorite organ subject.

When I learned about this fact early in my student days it was a depressing piece of evidence but when I realized humans especially doctors after graduation use only a fraction of the billions of nerve cells (that also to become rich) the chord of consolation was strung and that kept me involved in intellectual activities (like what I am doing now).

Nobody should get depressed about this piece of evidence.

All of us (including me) only use only a fraction of the billion of nerve cells in our entire life and the brain reserve is more than enough to cope even up to eighty years.

However, alcohol is a universal brain poison and atherosclerosis is a major disease of the brain (not only heart) and both should be avoided by the wise.

I perhaps would like to think that meditation activities (Metta Meditation) probably retard the aging process of the brain by releasing undesirable pent up energies.

My own view is instead of the early morning tot five minutes of Metta Meditation would suffice to arrest the aging process to a considerable degree in the adults after the age of 35 years.

There are many theories of aging and it is the least studied and least understood human process (physiological and not pathological) that affects the mankind and the scientific literature is known for its absence of concrete themes.

What is interesting is that the aging process is not seen in the wild and the less able individuals are eliminated by choice in nature.

It is only with the advancement of civilization that the aging process is selected by choice.

In the prehistoric times most men died around 35 years of age but women lived little longer (if maternal death in child birth did not intervene) to about 45 years.

It is interesting to note that the sturgeons that are killed for roe takes longtime to mature in fact they grow almost continuously and some large tortoises do not get old. Instead of decimating them for caviar and isinglass a look at their biological potential along with the tiny guppy fish (does not get old but continues to grow) is warranted. Please note that guppy fish die of a guppy disease (probably a mutant viral disease) only affecting them.

They do not get old like sturgeons.

The onset of aging process is later in women by 10 years but past the 45 years of age the aging process is rapidly accelerated and this may be related to lack of oestrogen hormone and the onset of androgenic effects on tissues with cell cycle events overtaking the longevities.

The biology and evolution have fashioned themselves in a manner that reproductive functions are intervened between termination of series of cell cycle events (aging process) and the beauty in nature is that there is a rich variety of species from unicellular to multi-cellular organisms.

There are four basic facets of aging. Firstly, the changes in aging are universal and occur in all members of the species. Secondly, aging is inherent and intrinsic to the individual and will occur even if all the environmental effects are removed. Thirdly, the aging is a progressive process, which starts slowly but gradually builds up its momentum in a cumulative fashion. Finally, the process is detrimental to the organism and results in shortening of life and ultimate death.

Programmed self destruction is mechanism behind aging process and is held in check in the genes from the onset of fertilization to maturity and from then onwards series of events in the cell cycle let go the killer mechanisms albeit in smaller doses. If this let go mechanism is not set in motion invariably a cancer develops and kills the organisms.

Hela cells that originated from a cancer, grow in the laboratory indefinitely and never die. However, in the laboratory it has been shown that normal animal cells can be propagated for a finite number of cell generations (cycles).

Every species have a finite number of life span and in mice there are about 30 cell cycles with a life span of 12 weeks. This is approximately equivalent to the growth of the human embryo but all the events from fertilization to maturity to death occur in only 3 months.

Old ovaries if transplanted into young mice are not rejuvenated but the young ovaries transplanted into old mice continue to function as young. This brings about an interesting preposition. A woman could have one of her ovaries removed and frozen in the laboratory when young and she could have this young ovary transplanted into her when she is old.

Future generations of medical research would bypass the basic human needs and go in this direction and this would certainly cost more than the present day beauty care.

For a man he can have one of his testicles deep frozen for future use and this is certainly desirable with the global warming since the testicles which needs a cooler atmosphere would love the super cooler state of a supermarket.

Unfortunately, for men the desirable effect of living longer is not possible with an aching heart and a young testicle since androgens have a deteriorating effect on long life.

I should wind up with some controversial issues lest this chapter has no meaning. If a women have a child at 15 years and then at 35 years (from a father who is 20 years and 40 years) what is the likely age of dying of these children who are both males.

I would say the firstborn would die at 80 years and the younger one by at 60 years.

I should state some scientific facts before elaborating on this poser. In the ovary germ cells undergo 20 to 30 mitotic cell divisions (normal division) in the first few months of embryonic life. By the age of three months (embryonic) these cells mature and start undergoing meiosis (reduction division) to female cells and remain in arrested maturation till puberty or menopause. The lengthy interval between first meiotic division and onset of ovulation up to 50 years later account for well documented increase in chromosomal abnormalities in the offspring of the older mothers. I believe in some way this is compounded by the aging process which affect accelerated aging process of the offspring having reached the adulthood.

In the male at puberty the primary sex cells that have undergone 30 cell divisions in embryogenesis enter into meiotic cell divisions (without any arrest as in the female) every 60 days and this is relatively a rapid rate (compared to some cancer cells) and by 50 years there over 1000 to 1500 mitotic divisions in all. The observation of dominant mutations is consistent with these rapid rates of cell divisions and old fathers having offspring with genetic abnormality.

The same argument above of aging hold true for male sex cells.

So coming to social conventions the youngest of the family (if he is a boy) getting all the benefits of the family heritage is not based on any scientific evidence. If family diseases are not intervening the older one would live longer perhaps more years than the younger one born

20 years later who would inherit more of genetic defects with an older generation of cells in his constitution even by chance coincidence. The biological cycle has some inherent mechanisms to check and balance the number in each successive generation. The death of a generation almost simultaneously within a short space of time is genetically designed and I call this the biological control of population.

This was evident in China before the introduction of one child family program. What I mean is even before the introduction of this family planning program the population was coming down in China due to increase in death rates of successive generations.

Another observation, I have made in some selected families is that the average age of death is tailing off, in fact coming down in Sri-Lankan context. This is probably because of late marriage. The average age of having children has gone up to 28 from 20 in the past half a century.

Is it that the children born to older couples are dying young?

That's another poser.

Another poser would make people revolt (especially the men) against my suggestions. A woman should marry a younger person at least by 5 years. The basic tenet is that women live longer and men die at younger age. Why women who are supposed to be liberated want to live without their husbands at their (most tender years) old age?

This is one situation where men have become wiser over generations and have scored a point even before the game started.

I mean the game of marriage.

I hope nobody takes me seriously and literally on these issues unless they are mad.

My point is we are said to be wise and scientific.

I have my reservations.

If one asks me the questions when (age) one should marry.

My prompt answer is when you feel like with a poser are you ready?

If he is wise and asks me a second question why did you marry?

After a pause and a pensive mood I would say one is allowed to make many minor mistakes in life but only one major.

This has worked for everybody.

Chapter 32
Biophysics

This is another word I would like to add to English, if it is not there already in existence, this time in Scientific World.

We had an Engineering Exhibition in the University, I do not know what was being exhibited there but my gut feeling says is that they are all big events for public consumption and certainly not miniature exhibits of biological nature.

I am pretty sure they did not talk about biophysics and nano-technology.

I have been against the Coal Power Project form its inception.

I have many reasons including acid rain and I have stated that Coal Marketing Board will be as bad as the Paddy Marketing Board of yesteryear with built in corruption. Suffice is to say that most of the Coal would wind up in undesirable places and the scale of corruption would be covered by black soot.

How black soot will effect our children's chest physiology and sporting outcomes are a different kettle of fish. I have voiced these concerns with scientific data but they are pushed under the carpet of power politics.

One pertinent question our energy planners not addressed is that these coal power plants are operational 24 hours a day and the interruption of nocturnal terrestrial cooling effect that is necessary for air circulation and water condensation and the night rain pattern that we used enjoy that invariably purified the air of soot, dirt and chemicals is going to be disturbed due to constantly warm air that is not circulating.

The effect of this is going to be phenomenal.

As long as the rain is there even infrequently and with disturbed patterns then there is some reclamation.

But think of the scenario for some reason or the other if the rain fails for a period of two years with drought.

Then we have to be using more and more coal power to generate power, the conditions are conducive for catastrophic events.

The nocturnal cooling effect is negated for a prolonged period of time and the pollution it creates is of clinical scale.

We see this pattern in Kandy even without the warming effect discussed above and that is why Kandy is the most polluted city of the country. When the rain fails the nocturnal cooling draft fails and the pollutants remain for considerable length of time in the lower atmosphere.

We have now cleared our rain forest to less (33% is the critical value) than the sustainable level for natural rain and ecological balance.

We can see the effect of acid rain in less than 5 years after a period of prolonged drought.

I will talk a little about the miniature life of biological nature that can mop up some of the

CO_2 created by the mega projects like coal power plants.

There are more than thousands of other chemicals that are emitted which cannot be made safe and some of them can cause cancer.

These are mega events and banning cigarettes has no ameliorating effects once these chemicals are in air and in circulation and we have to breath them 24 hours a day.

The algae I had been interested in are tiny but common in nature and their contribution can be very significant if properly harnessed but the experiments done on them are very few an far between.

I gather in Germany they are working on them to mop up CO_2 emitted by coal power units.

What ever big or small engineering feats it may be, whether it is motor car or a rocket, no engineering feat can bring about more than 33% efficiency or throughput.

In biological systems this is scaled down to less than 3 to 4% and even though the process is very slow but by their shear (number of cells) numbers and their shorter multiplication cycles (multiplication by cell division), the process of conversion of CO_2 can be substantial and can reverse the potential rise and make it to drop.

Problem with engineering feats is that they are not cost effective or efficient.

The rate of conversion is fast and the rate of production of pollutants and poisons are also fast.

This is what I used to argue with my fellow engineering students at the campus many moons ago and also asked why they used heavy metal to built cars (that also not stainless steel) that is subjected to resistance or impedance or inertia of motion and why not use a metal like aluminum used in aircrafts.

They will give some answers and I will give counter arguments while playing bridge (which I learnt from them) and this goes on with new themes added.

Years went by we parted and departed from ivory towers we built and went into real world and domestics. It is sad now I cannot find somebody to talk ten words in analytical English on a scientific topic or discussion.

Even, retired dons, talking and taking partisan in bizarre politics in their twilight years sadden me most. I am not saying that they are not entitled to voice their opinions. With their reputation if they make mistakes their ardent followers are left astray.

What has happened to their scientific inquiry and what has happened to their inbuilt instincts.

We were fed a good and healthy dose of inquiry and free discussion then and now there is paucity of inquiry and query, question or rationalistic views about world around us living and inanimate.

I dearly remember Prof. Osmond Jayaratne and his investigation into lightening. There was Prof. Rnjith Ruberu with beautiful exposition of Botany and Biology.

I am not talking about Chintanaya Professors with mathematical talents.

I never talked about biophysics then since all medical stuff including biochemistry was boring for me.

Now that I am somewhat of an independent thinker my argument of biophysics start with a simple relationship (not Newton's equal and opposite reaction).

This is where I have element of disagreement with Buddhist practices, too.

Why only animals are sacred?

Why not other living things, the plants that all animals depend directly and indirectly?

To me plants are sacred too.

Without plant life the world cannot exist.

This is where my thinking is at variance with even Buddhist tenet.

Why there is only a law for animals?

Is it only animals with brains or nervous system that is important to Buddhists?

I have not found an answer for this in Buddhist scriptures.

So I am entitled to my own analysis in a biological sense.

So my simple theoretical aspect of biological relationships especially the biophysics is that there is infinite relationship one to one and one to many in every biological cell or biological system.

There is no distinction of plant life and animal life whatever the form it may be.

They are all life forms in a constant web of actions and reactions.

As far as we know they can be seen with definite evidence only on this planet.

This planet is the only planet we know of and I am not extrapolating to any other heavenly objects with life or without life in this universe.

Ever since the life began in this planet this biological or biophysical relationship continuously existed without a break.

There is either positive or negative relationship but this relationship is such that it was never to destroy or eliminate one or the other.

Continuity of life existed uninterrupted in spite of major physical catastrophes.

If this would not have being the case various new forms of life would not have emerged in evolution.

There was competition but never extermination by living thing by living things.

That is my bona fide or the compelling argument.

All life forms are important including mosquitoes whether they carry infectious agents or not.

We cannot argue that world should be without insects including mosquitoes.

There my argument and tenet converge with Buddhism.

All life forms be happy!

It is the physical catastrophes that eliminated species lock stock and barrel but never simple biological competition.

This theoretical aspect had very interesting convergence when I saw my fish dying in the open including the Guppies.

First I thought it was the algal bloom and the lack of oxygen.

It was the very simple and straight forward plausible explanation I could give.

But when this happened again and when I checked with the internal digital temperature it was not strange coincidence that the temperature was above 95 degrees Fahrenheit.

It was the global warming (atmospheric temperature) that killed the fishes, other factors were contributory.

Interestingly high atmospheric temperature encourages mosquito breeding.

Every ton of coal (when it is burnt 24 hours a day) we burn will increase the number of mosquitoes by billions!

Single physical change, the temperature alone can kill animals.

Humans can die of heat stroke and why not fish.

There is a protein called heat shock protein even in bacteria and our cells also have it in a different form and all animals do have this protein to overcome stress.

Heat is a stress.

That system can easily be damaged.

Yes most of the fish who are usually kept indoors especially the mollies cannot live above 86 degrees.

It was only algal bloom and lack oxygen alone then some fish could come up and breath at the surface or produce young ones before dying but this did not happen.

The biological relationships I propose in theory could not keep pace with the adverse temperature which is partly man made.

There was no time for evolutionary adaptation.

The other relationship is human population expansion.

I believe we have come to the optimum population of 6 billions.

The earth cannot sustain 9 billions.

This extra 3 billion will upset the balance of biophysical relationship that come into equilibrium in the microscopic levels.

The mega level is man made due to his megalomaniac beahviour.

Earth cannot sustain its biophysical relationships with the rate at which the human population is expanding and consuming all the limited resources.

Biophysical barrier will break down soon.

Then calamities after calamities would occur.

Gloom, bloom and doom unless we arrest the population expansion and the rate of use of easily available energy resources.

First we have to arrest the population growth.

Second we have to have a food security.

Third we have to prevent the made made causes of global warming.

Even in this country the priority is on energy and its use and its expansion and not conservation.

We got our priorities wrong.

We will loose all the biodiversity and elephants in no time.

We do not have neither the master plan nor the vision.

We do not have philosophers.

We do not have scientists of international caliber.

We only have politicians of various shades and colours but without any distinction.

Chapter 33

Cycle of Events

Humans and humanoids (a subspecies evolving out of global economic concepts with overpowering and obsessive desire to consume, usually white in colour but there are other shades including brown) think that they are rational beings but a simple observation made me to think otherwise.

I was traveling in a bus from Peradeniya to Kandy in the morning. I had nothing else to do, I decided to calculate the speed of the bus. It took 45 minutes and the speed of the bus was 8 km/hour. It was an amazing speed for a person who loved driving at a speed of 100 km/hour on a highway abroad.

Being myself bit inquisitive I wanted to find the average speed of a 'push' bike.

At a leisurely pace it is 12 to 20 km/ hour.

So if I had a speed of 12 km/hour I could have gone to Kandy and come back in an hour the most, with a pint of milk and the Sunday Paper.

Incidentally the pint of milk is roughly the energy and water requirement of my journey and the calcium is for my aging back.

To travel 12 kilometers on a bike the cyclist needs about 300 kcal. (It is sad but true that there are only two places where I can buy fresh unsweetened milk. Being a Dental Health Convert, I preach the value of fresh milk as against the sweetened milk.)

I put a little teaser to my aging gray (brain) cells. To travel the same distance an average car consumes staggering 48,000 kcal (200 MJ) and a gallon of petrol. A colossal waste of energy just to pick the Sunday Paper (by car) which is thrown away the next day. (The amount of additional Oxygen consumed by me is about 3.5 moles and the amount of Carbon Dioxide exhaled is about 2.5 moles if I cycled).

The amount of carbon dioxide produced by the car is staggering 400 moles.

The number of trees left in the city is not sufficient to absorb this surplus of carbon dioxide (by photosynthesis) produced by a single car journey (even if they live longer than me).

Coming back to humanoids rationale, the car which was invented for the purpose of speed and efficiency, is the biggest hindrance to efficiency due to the over use and congestion in the cities. The city planners and think tanks now work harder to built new roads and subways. This has been found counterproductive in M25 motorway round London. In no time congestion builds up again and we are back to square one and to the drawing board.

That is for our economists and Television Pundits to ponder.

The bigger the project still bigger is the disaster unforeseen.

There are about 500 million cars and 800 million bicycles in the world. The love of the car is phenomenal even though most of the people will never own car in their life time.

It is a status symbol.

The cycle is seen as a symbol of poverty even, in Africa.

In China there are about 300 million cycles and a regime that favours cycling work force, now has taken a U-turn like the New Labour Party in England. The country that boasted production of over 36 million bicycles is now producing 30 million (20% less) and the rate of increase of production of cars which was 4% in 1996 has reached 10%.

100 bicycles can be produced for the cost of building a car. In other words the bicycle is cheap to produce but the trend in China is in the opposite direction.

Probably the children born to single child families (probably single parent too) are either ambidextrous or they were not taught which is left and which is right like the New Labour Party in UK.

In our times bicycle was a popular mode of transport.

Only problem was how to prevent Gajayas (friends) taking the bike away in the middle of the night.

However, medical students (some of our professors came to work in scooters) had a penchant for scooters.

Apart from health benefits, its convenience, cycling was a pleasurable pastime in our days.

Oxford university somewhat similar to Peradeniya is known for its bicycle commuters even today. Now only a few minor employees use this mode of transport regularly and the time has come to question our wisdom.

Unfortunately, with traffic congestion and so many three wheelers on the road cycling has become a potential hazard to young undergraduates. This has become a reason for the decline in bicycles in the campus.

Bicycle perceived as a low status (artificial) symbol is used as an exercise (stationary though) machine in sports facilities defeating its rightful place as the poor man's friend.

Are we proving to ourself we are prudent in our choices?

The environmental pollution (air and noise) contributed by motor vehicles is accepted not as an evil but as a champion of development.

Ironically, two way catalytic converters built to reduce carbon monoxide emissions and smog do not reduce oxides of nitrogen, (which is 300 times more potent in ozone depletion) only work on lead-free fuel.

Catalytic converters do not work on diesel engines. The irony in India and Sri-Lanka is that there is higher tax on petrol (which is environmentally less harmful) and subsidies on diesel (which promotes the use of diesel engines in preference to petrol engines).

Cleaner and the environmentally friendly people pay taxes for the benevolence of the polluters.

Coal power plants recently commissioned would in a matter of 50 years destroy the remaining rain forest and our grand children would be pedaling (with a mask on) bicycles at school to power the computer they are using (hydroelectric power non existing due to drought; sipping water drop by drop from a bottle to prevent dehydration. The water is a scarce commodity).

My grand child would be so advanced in his skills that not only he can ride, type, eat and drink in a stride, while also taking part in live chat with uncle Sam in America and uncle Albert in England, asking for a little loan for an electric bicycle (by then one of the most expensive presents for an average Sri-Lankan).

Shanghai one of the most polluted cities of the world would have more deaths related to respiratory illness if the present trend of increase of motor vehicles continues. It looks as if the policy makers of China let its comrades die (simply an altruistic urge) prematurely so that the young and strong can survive.

I thought and believed that China produced philosophers and rational economists. Probably may have been killed in the cultural revolution.

Interestingly, United States has more bicycles than in China but only 2.5% of Americans use it for commuting to work.

Instead they use it for sport and recreation.

So in America there are only 2.5% rational and practical men in spite of so many universities.

So how come, the American Colleges that are coming up in Sri-Lanka produce rational economists and scientists of Sri-Lankan origin?

It beats me the logic of going to an American College for Graduation if only 2.5% of them use their brains after graduation.

But there is an outstanding American called Lance Armstrong (I do not know to which college he went. I suggest any Sri-Lankan who is bright enough must enter that college in Texas) who made the Tour de France, the "Lance de France".

Battling illness and cancer with courage and determination he won the Tour de France with grace as a true world champion.

He is one who is worth emulating, as a model of a rational man, if his doping can be forgiven.

French people deserve a special thanks for popularizing cycling as a way of life and sport.

I want the American and French Embassy to recruit our retiring political (unfortunately they never consider retiring as a healthy pastime) giants and train them in cycling either in America or in France so that they lose weight and live ever so happily thereafter.

They are overweight since they have undisclosed number of official vehicles to transfer from kitchen cabinet to the potato hotpot.

I have an e-question to the current US President.

How come an affluent country like US with high growth rates tend to be the most polluted and afflicted by public squalor at an astonishing speed, both at the same time?

Should Sri-Lanka follow the same path?

Chapter 34

Epilogue

Spacesuit and its Occupant

This is an idea I hit upon by reading a blog writing of a experienced civil pilot not a fighter pilot.

Have you ever thought of the 50 things that a spacesuit occupant in space won't share with his body soul?

You probably have not but I was one who was very much interested in this in my school days and in early days as a medical student.

I cannot remember what I wrote then but this is an attempt to revisit and revise some of those physiological constraints not in any particular order or in order of merits.

Suffice is to say I get a sickly feeling when I think of the space (occupied especially by alien elements) and spacesuit.

Imagine yourself trapped in a escalator without illumination and the computer circuit controlling its up and down movement had gone haywire and it is going up and down in an erratic fashion.

That is a the feeling I get the moment I put a spacesuit on for travel.

That is one thing you must consider when paying for, an enormous amount for a single trip in space.

Is it worth the experience and the money?

Probably not but having said that I have tremendously high regard for those guys who trained for years to go to space. They are a dedicated lot and give them the due respect they deserve if you happen to meet anyone of them.

They were the human guinea pigs on space.

I often wonder how many times they felt sick and vertiginous even in their sleep.

Probably many, many times and uncountable and that is the feeling I get if I am invited to wear a spacesuit and come hither for a go.

I will list the feeling inside my head with little imagination and some understanding of my own physiology if not of another being.

1. I hate the space constraint.

This is the feeling one gets if one has to stay in a tiny hotel room in Singapore overnight due to some delay, cancellation or transit. I have had that feeling once or twice before, traveling by cheap air flights. When you fly on a good aircraft you may not get this inconvenience but surely on a budget air.

2. I hate heights.

Imagine you are in a hotel overnight on the 21st floor room due to flight cancellation. You are well away from a fire exit and there is a blackout and fire drill. You don't have a pen torch. If you are one floor above you can think of jumping out and breaking your legs but not from the 21st floor. That is why rooms are cheap as you go up. Please pay a good some and ask for a room down below. It is better even now if you go to Colombo taking a pen torch with you with these high rising development projects.

Born to this earth with feet firmly grounded and ample space to breath pristine air (not now even in Kandy) as an embodiment, getting into a spacesuit is the luxury I do not want to avail myself not even in my incarnation.

3. Now about the daily routines I enjoy.

Sleep to begin with.

I think I can manage sleep upside down in space inside a spacesuit since there is nothing else I can do there except dreaming coming home. I can do this since I have learned how to sleep standing on an express bus plying from Kandy to Colombo on a Monday morning. I believe all Sri-Lankans are good at this.

Only if you do not have money in your back pocket. There are plenty of pickpockets in this country including politicians who pick our vote without our knowledge.

They are called pickvotters or even better pickpotters (stuffing the ballot boxes).

These two are new words, I have coined for the Oxford Dictionary with the local elections due.

4. What about food.

I won't enjoy the high calorie, high protein dehydrated food fads of space travelers especially if they are floating about around you not as sandwiches and not placed on a plate with a well laid out table.

My worry is not the quality of food but how I to partake them in a more sociable way.

Not empty them to my mouth from one paper carton to another.

5. Coming to spirits (if they are allowed like a commercial flight) and drinks.

When I suck (not drink them) a little, I want them to stay a while in the mouth and oesophagus and stomach and not go flushing down like a vacuum cleaner on full throttle to the colon in one go.

6. After meal I want to brush my teeth as my good dental friends tell me with a tooth brush floating in air and the toothpaste all over the face with me trying to reach as far as it goes to the third molar.

7. That also I can manage but how about a quick spend a penny in the loo with my prostrate pushing hard on the correct track inside but the squirt getting between my spacesuit and the underwear.

That is my major worry since I will never master my physiology how ever much I train on earth and mid air.

8. Then the master job of course I have decided one last one here and never in the shuttle till I come home and take some constipating medicine once a week for six weeks before the scheduled departure.

I do not want my smelly secrets floating in air and taking pictures of me in flight.

No thank you.

9. Last but not least I fear the algae and the fungi I have been accustomed on earth and living with me with mutual understanding all along my life for years taking advantage of the flight and growing all over me.

In nails, wind pipe, mouth and all orifices on my privacies.

10. Last but not least, I love scratching my skin, just for fun and any other accessible point from my crown to the rump.

With these fungi floating around and waiting for a breach, I won't be able enjoy that luxury.

11. As for the rubbish I collect on flight no problem.

We are trained to drop at any advantage point in the town and the Municipality never clean them up. I just open the window and drop them down when we are centering above Sri-Lanka with a note stating; "Coming from space shuttle in orbit no valuables dropped but destined for Sri-Lankans, war heroes included".

This is why when President Obama invited me for a flight in space, I refused and gave over 100 volunteers from our parliament elected and wanting to get elected.

He of course refused nay to all parliamentarians after the Health Bill was Bailed out with a American Donald Duckbill.

Authors Note

This book is intended for anybody who is interested in science.

For some reason teaching science has become boring for many reasons. Teachers try to fit into a curriculum whereas children are developing. One need to understand their developmental status and their ability to integrate science from fiction.

It has to be flexible and innovative.

Loading them with factual knowledge and equations kills their interest if they are not applicable to their way of life.

I hope, I have stimulated the reader to take some interest in science as a hobby and way of life.

Unfortunately, the poor teaching of science is global.

Asokaplus

Asokaplus